Statistical Data Analysis

T0201853

Statistical Data Analysis

GLEN COWAN

University of Siegen

CLARENDON PRESS · OXFORD

1998

This book has been printed digitally in order to ensure its continuing availability

OXFORD
UNIVERSITY PRESS

Great Clarendon Street, Oxford OX2 6DP

Oxford University Press is a department of the University of Oxford.
It furthers the University's objective of excellence in research, scholarship,
and education by publishing worldwide in

Oxford New York

Auckland Bangkok Buenos Aires Cape Town Chennai
Dar es Salaam Delhi Hong Kong Istanbul Karachi Kolkata
Kuala Lumpur Madrid Melbourne Mexico City Mumbai Nairobi
São Paulo Shanghai Singapore Taipei Tokyo Toronto

with an associated company in Berlin

Oxford is a registered trade mark of Oxford University Press
in the UK and in certain other countries

Published in the United States
by Oxford University Press Inc., New York

© Glen Cowan, 1998

The moral rights of the author have been asserted
Database right Oxford University Press (maker)

Reprinted 2002

A catalogue record for this book is available from the British Library

Library of Congress Cataloging in Publication Data
(Data available)

ISBN 0-19-850156-0 (Hbk)
ISBN 0-19-850155-2 (Pbk)

Preface

The following book is a guide to the practical application of statistics in data analysis as typically encountered in the physical sciences, and in particular in high energy particle physics. Students entering this field do not usually go through a formal course in probability and statistics, despite having been exposed to many other advanced mathematical techniques. Statistical methods are invariably needed, however, in order to extract meaningful information from experimental data.

The book originally developed out of work with graduate students at the European Organization for Nuclear Research (CERN). It is primarily aimed at graduate or advanced undergraduate students in the physical sciences, especially those engaged in research or laboratory courses which involve data analysis. A number of the methods are widely used but less widely understood, and it is therefore hoped that more advanced researchers will also be able to profit from the material. Although most of the examples come from high energy particle physics, an attempt has been made to present the material in a reasonably general way so that the book can be useful to people in most branches of physics and astronomy.

It is assumed that the reader has an understanding of linear algebra, multivariable calculus and some knowledge of complex analysis. No prior knowledge of probability and statistics, however, is assumed. Roughly speaking, the present book is somewhat less theoretically oriented than that of Eadie *et al.* [Ead71], and somewhat more so than those of Lyons [Lyo86] and Barlow [Bar89].

The first part of the book, Chapters 1 through 8, covers basic concepts of probability and random variables, Monte Carlo techniques, statistical tests, and methods of parameter estimation. The concept of probability plays, of course, a fundamental role. In addition to its interpretation as a relative frequency as used in classical statistics, the Bayesian approach using subjective probability is discussed as well. Although the frequency interpretation tends to dominate in most of the commonly applied methods, it was felt that certain applications can be better handled with Bayesian statistics, and that a brief discussion of this approach was therefore justified.

The last three chapters are somewhat more advanced than those preceding. Chapter 9 covers interval estimation, including the setting of limits on parameters. The characteristic function is introduced in Chapter 10 and used to derive a number of results which are stated without proof earlier in the book. Finally, Chapter 11 covers the problem of unfolding, i.e. the correcting of distributions for effects of measurement errors. This topic in particular is somewhat specialized, but since it is not treated in many other books it was felt that a discussion of the concepts would be found useful.

An attempt has been made to present the most important concepts and tools in a manageably short space. As a consequence, many results are given without proof and the reader is often referred to the literature for more detailed explanations. It is thus considerably more compact than several other works on similar topics, e.g. those by Brandt [Bra92] and Frodeson *et al.* [Fro79]. Most chapters employ concepts introduced in previous ones. Since the book is relatively short, however, it is hoped that readers will look at least briefly at the earlier chapters before skipping to the topic needed. A possible exception is Chapter 4 on statistical tests; this could by skipped without a serious loss of continuity by those mainly interested in parameter estimation.

The choice of and relative weights given to the various topics reflect the type of analysis usually encountered in particle physics. Here the data usually consist of a set of observed events, e.g. particle collisions or decays, as opposed to the data of a radio astronomer, who deals with a signal measured as a function of time. The topic of time series analysis is therefore omitted, as is analysis of variance. The important topic of numerical minimization is not treated, since computer routines that perform this task are widely available in program libraries.

At various points in the book, reference is made to the CERN program library (CERNLIB) [CER97], as this is the collection of computer sofware most accessible to particle physicists. The short tables of values included in the book have been computed using CERNLIB routines. Other useful sources of statistics software include the program libraries provided with the books by Press *et al.* [Pre92] and Brandt [Bra92].

Part of the material here was presented as a half-semester course at the University of Siegen in 1995. Given the topics added since then, most of the book could be covered in 30 one-hour lectures. Although no exercises are included, an evolving set of problems and additional related material can be found on the book's World Wide Web site. The link to this site can be located via the catalogue of the Oxford University Press home page at:

http://www.oup.co.uk/

The reader interested in practicing the techniques of this book is encouraged to implement the examples on a computer. By modifying the various parameters and the input data, one can gain experience with the methods presented. This is particularly instructive in conjunction with the Monte Carlo method (Chapter 3), which allows one to generate simulated data sets with known properties. These can then be used as input to test the various statistical techniques.

Thanks are due above all to Sönke Adlung of Oxford University Press for encouraging me to write this book as well as for his many suggestions on its content. In addition I am grateful to Professors Sigmund Brandt and Claus Grupen of the University of Siegen for their support of this project and their feedback on the text. Significant improvements were suggested by Robert Cousins, as well as by many of my colleagues in the ALEPH collaboration, including Klaus Affholderbach, Paul Bright-Thomas, Volker Büscher, Günther Dissertori, Ian Knowles, Ramon Miquel, Ian Tomalin, Stefan Schael, Michael Schmelling and Steven Wasserbaech. Last but not least I would like to thank my wife Cheryl for her patient support.

Geneva
August 1997   G.D.C.

Contents

Notation

Throughout this book, $\log x$ refers to the natural (base e) logarithm. Frequent use is made of the Kronecker delta symbol,

$$\delta_{ij} = \begin{cases} 1 & i = j, \\ 0 & \text{otherwise.} \end{cases}$$

Although there are occasional exceptions, an attempt has been made to adhere to the following notational conventions.

$P(A)$	probability of A
$P(A\|B)$	conditional probability of A given B
$x, y, t, \ldots$	continuous (scalar) random variables
$\mathbf{x} = (x_1, \ldots, x_n)$	vector of random variables
$f(x), g(x), \ldots$	probability densities for x
$F(x), G(x), \ldots$	cumulative distributions corresponding to p.d.f.s $f(x), g(x), \ldots$
$f(x, y)$	joint probability density for x, y
$f(x\|y)$	conditional probability density for x given y
n	discrete (e.g. integer) random variable
ν	expectation value of n (often, Greek letter = expectation value of corresponding Latin letter)
$\theta, \alpha, \beta, \xi, \tau, \ldots$	(scalar) parameters
$\boldsymbol{\theta} = (\theta_1, \ldots, \theta_n)$	vector of parameters
$f(x; \theta)$	probability density of x, depending on the parameter θ
$E[x]$	expectation value of x (often denoted by μ)
$V[x]$	variance of x (often denoted by σ^2)
$\text{cov}[x_i, x_j]$	covariance of x_i, x_j (often denoted by matrices $V_{ij}, U_{ij}, \ldots$)
$\hat{\theta}, \widehat{\sigma^2}, \ldots$	estimators for $\theta, \sigma^2, \ldots$

$\hat{\theta}_{\text{obs}}$	an observed value of the estimator $\hat{\theta}$
$\overline{x}$	arithmetic mean of a sample $x_1, \ldots, x_n$
μ'_n	nth algebraic moment
μ_n	nth central moment
x_α	α-point, quantile of order α
$\varphi(x)$	standard Gaussian p.d.f.
$\Phi(x)$	cumulative distribution of the standard Gaussian
$\phi(k)$	characteristic function
$L(\theta), L(\mathbf{x}|\theta)$	likelihood function
$\pi(\theta)$	prior probability density
$p(\theta|\mathbf{x})$	posterior probability density for θ given data $\mathbf{x}$

1
Fundamental concepts

1.1 Probability and random variables

The aim of this book is to present the most important concepts and methods of statistical data analysis. A central concept is that of uncertainty, which can manifest itself in a number of different ways. For example, one is often faced with a situation where the outcome of a measurement varies unpredictably upon repetition of the experiment. Such behavior can result from errors related to the measuring device, or it could be the consequence of a more fundamental (e.g. quantum mechanical) unpredictability of the system. The uncertainty might stem from various undetermined factors which in principle could be known but in fact are not. A characteristic of a system is said to be **random** when it is not known or cannot be predicted with complete certainty.

The degree of randomness can be quantified with the concept of **probability**. The mathematical theory of probability has a history dating back at least to the 17th century, and several different definitions of probability have been developed. We will use the definition in terms of set theory as formulated in 1933 by Kolmogorov [Kol33]. Consider a set S called the **sample space** consisting of a certain number of elements, the interpretation of which is left open for the moment. To each subset A of S one assigns a real number $P(A)$ called a probability, defined by the following three axioms:[1]

(1) For every subset A in S, $P(A) \geq 0$.
(2) For any two subsets A and B that are disjoint (i.e. mutually exclusive, $A \cap B = \emptyset$) the probability assigned to the union of A and B is the sum of the two corresponding probabilities, $P(A \cup B) = P(A) + P(B)$.
(3) The probability assigned to the sample space is one, $P(S) = 1$.

From these axioms further properties of probability functions can be derived, e.g.

---

[1]The axioms here are somewhat simplified with respect to those found in more rigorous texts, such as [Gri92], but are sufficient for our purposes. More precisely, the set of subsets to which probabilities are assigned must constitute a so-called σ-field.

$$P(\overline{A}) = 1 - P(A) \text{ where } \overline{A} \text{ is the complement of } A$$
$$P(A \cup \overline{A}) = 1$$
$$0 \le P(A) \le 1$$
$$P(\emptyset) = 0 \tag{1.1}$$
$$\text{if } A \subset B, \text{ then } P(A) \le P(B)$$
$$P(A \cup B) = P(A) + P(B) - P(A \cap B).$$

For proofs and further properties see e.g. [Bra92, Gri86, Gri92].

A variable that takes on a specific value for each element of the set S is called **a random variable**. The individual elements may each be characterized by several quantities, in which case the random variable is a multicomponent vector.

Suppose one has a sample space S which contains subsets A and B. Provided $P(B) \neq 0$, one defines the **conditional probability** $P(A|B)$ (read P of A given B) as

$$P(A|B) = \frac{P(A \cap B)}{P(B)}. \tag{1.2}$$

Figure 1.1 shows the relationship between the sets A, B and S. One can easily show that conditional probabilities themselves satisfy the axioms of probability. Note that the usual probability $P(A)$ can be regarded as the conditional probability for A given S: $P(A) = P(A|S)$.

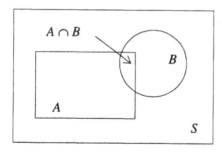

Fig. 1.1 Relationship between the sets A, B and S in the definition of conditional probability.

Two subsets A and B are said to be **independent** if

$$P(A \cap B) = P(A)\, P(B). \tag{1.3}$$

For A and B independent, it follows from the definition of conditional probability that $P(A|B) = P(A)$ and $P(B|A) = P(B)$. (Do not confuse independent subsets according to (1.3) with disjoint subsets, i.e. $A \cap B = \emptyset$.)

From the definition of conditional probability one also has the probability of B given A (assuming $P(A) \neq 0$),

$$P(B|A) = \frac{P(B \cap A)}{P(A)}. \tag{1.4}$$

Since $A \cap B$ is the same as $B \cap A$, by combining equations (1.2) and (1.4) one has

$$P(B \cap A) = P(A|B)\, P(B) = P(B|A)\, P(A), \qquad (1.5)$$

or

$$P(A|B) = \frac{P(B|A)\, P(A)}{P(B)}. \qquad (1.6)$$

Equation (1.6), which relates the conditional probabilities $P(A|B)$ and $P(B|A)$, is called **Bayes' theorem** [Bay63].

Suppose the sample space S can be broken into disjoint subsets A_i, i.e. $S = \cup_i A_i$ with $A_i \cap A_j = \emptyset$ for $i \neq j$. Assume further that $P(A_i) \neq 0$ for all i. An arbitrary subset B can be expressed as $B = B \cap S = B \cap (\cup_i A_i) = \cup_i (B \cap A_i)$. Since the subsets $B \cap A_i$ are disjoint, their probabilities add, giving

$$
\begin{aligned}
P(B) &= P(\cup_i (B \cap A_i)) = \sum_i P(B \cap A_i) \\
&= \sum_i P(B|A_i) P(A_i).
\end{aligned}
\qquad (1.7)
$$

The last line comes from the definition (1.4) for the case $A = A_i$. Equation (1.7) is called the **law of total probability**. It is useful, for example, if one can break the sample space into subsets A_i for which the probabilities are easy to calculate. It is often combined with Bayes' theorem (1.6) to give

$$P(A|B) = \frac{P(B|A)\, P(A)}{\sum_i P(B|A_i) P(A_i)}. \qquad (1.8)$$

Here A can be any subset of S, including, for example, one of the A_i.

As an example, consider a disease which is known to be carried by 0.1% of the population, i.e. the **prior probabilities** to have the disease or not are

$$P(\text{disease}) = 0.001,$$
$$P(\text{no disease}) = 0.999.$$

A test is developed which yields a positive result with a probability of 98% given that the person carries the disease, i.e.

$$P(+|\text{disease}) = 0.98,$$
$$P(-|\text{disease}) = 0.02.$$

Suppose there is also a 3% probability, however, to obtain a positive result for a person without the disease,

$$P(+|\text{no disease}) = 0.03,$$
$$P(-|\text{no disease}) = 0.97.$$

What is the probability that you have the disease if your test result is positive? According to Bayes' theorem (in the form of equation (1.8)) this is given by

$$
\begin{aligned}
P(\text{disease}|+) \;&=\; \frac{P(+|\text{disease})\,P(\text{disease})}{P(+|\text{disease})\,P(\text{disease}) \,+\, P(+|\text{no disease})\,P(\text{no disease})} \\[2mm]
&=\; \frac{0.98 \times 0.001}{0.98 \times 0.001 + 0.03 \times 0.999} \\[2mm]
&=\; 0.032.
\end{aligned}
$$

The probability that you have the disease given a positive test result is only 3.2%. This may be surprising, since the probability of having a wrong result is only 2% if you carry the disease and 3% if you do not. But the prior probability is very low, 0.1%, which leads to a **posterior probability** of only 3.2%. An important point that we have skipped over is what it means when we say $P(\text{disease}|+) = 0.032$, i.e. how exactly the probability should be interpreted. This question is examined in the next section.

1.2 Interpretation of probability

Although any function satisfying the axioms above can be called by definition a probability function, one must still specify how to interpret the elements of the sample space and how to assign and interpret the probability values. There are two main interpretations of probability commonly used in data analysis. The most important is that of **relative frequency**, used among other things for assigning statistical errors to measurements. Another interpretation called **subjective** probability is also used, e.g. to quantify systematic uncertainties. These two interpretations are described in more detail below.

1.2.1 Probability as a relative frequency

In data analysis, probability is most commonly interpreted as a **limiting relative frequency**. Here the elements of the set S correspond to the possible outcomes of a measurement, assumed to be (at least hypothetically) repeatable. A subset A of S corresponds to the occurrence of any of the outcomes in the subset. Such a subset is called an **event**, which is said to occur if the outcome of a measurement is in the subset.

A subset of S consisting of only one element denotes a single **elementary outcome**. One assigns for the probability of an elementary outcome A the fraction of times that A occurs in the limit that the measurement is repeated an infinite number of times:

$$P(A) = \lim_{n \to \infty} \frac{\text{number of occurrences of outcome } A \text{ in } n \text{ measurements}}{n}. \quad (1.9)$$

The probabilities for the occurrence of any one of several outcomes (i.e. for a non-elementary subset A) are determined from those for individual outcomes by the addition rule given in the axioms of probability. These correspond in turn to relative frequencies of occurrence.

The relative frequency interpretation is consistent with the axioms of probability, since the fraction of occurrences is always greater than or equal to zero, the frequency of any out of a disjoint set of outcomes is the sum of the individual frequencies, and the measurement must by definition yield some outcome (i.e. $P(S) = 1$). The conditional probability $P(A|B)$ is thus the number of cases where both A and B occur divided by the number of cases in which B occurs, regardless of whether A occurs. That is, $P(A|B)$ gives the frequency of A with the subset B taken as the sample space.

Clearly the probabilities based on such a model can never be determined experimentally with perfect precision. The basic tasks of **classical statistics** are to estimate the probabilities (assumed to have some definite but unknown values) given a finite amount of experimental data, and to test to what extent a particular model or theory that predicts probabilities is compatible with the observed data.

The relative frequency interpretation is straightforward when studying physical laws, which are assumed to act the same way in repeated experiments. The validity of the assigned probability values can be experimentally tested. This point of view is appropriate, for example, in particle physics, where repeated collisions of particles constitute repetitions of an experiment. The concept of relative frequency is more problematic for unique phenomena such as the big bang. Here one can attempt to rescue the frequency interpretation by imagining a large number of similar universes, in some fraction of which a certain event occurs. Since, however, this is not even in principle realizable, the frequency here must be considered as a mental construct to assist in expressing a degree of belief about the single universe in which we live.

The frequency interpretation is the approach usually taken in standard texts on probability and statistics, such as those of Fisher [Fis90], Stuart and Ord [Stu91] and Cramér [Cra46]. The philosophy of probability as a frequency is discussed in the books by von Mises [Mis51, Mis64].

1.2.2   Subjective probability

Another probability interpretation is that of **subjective** (also called **Bayesian**) probability. Here the elements of the sample space correspond to **hypotheses** or **propositions**, i.e. statements that are either true or false. (When using subjective probability the sample space is often called the hypothesis space.) One interprets the probability associated with a hypothesis as a measure of degree of belief:

$$P(A) = \text{degree of belief that hypothesis } A \text{ is true.} \quad (1.10)$$

The sample space S must be constructed such that the elementary hypotheses are mutually exclusive, i.e. only one of them is true. A subset consisting of more than one hypothesis is true if any of the hypotheses in the subset is true. That is, the union of sets corresponds to the Boolean OR operation and the intersection corresponds to AND. One of the hypotheses must necessarily be true, i.e. $P(S) = 1$.

The statement that a measurement will yield a given outcome a certain fraction of the time can be regarded as a hypothesis, so the framework of subjective probability includes the relative frequency interpretation. In addition, however, subjective probability can be associated with, for example, the value of an unknown constant; this reflects one's confidence that its value lies in a certain fixed interval. A probability for an unknown constant is not meaningful with the frequency interpretation, since if we repeat an experiment depending on a physical parameter whose exact value is not certain (e.g. the mass of the electron), then its value is either never or always in a given fixed interval. The corresponding probability would be either zero or one, but we do not know which. With subjective probability, however, a probability of 95% that the electron mass is contained in a given interval is a reflection of one's state of knowledge.

The use of subjective probability is closely related to Bayes' theorem and forms the basis of **Bayesian** (as opposed to classical) statistics. The subset A appearing in Bayes' theorem (equation (1.6)) can be interpreted as the hypothesis that a certain theory is true, and the subset B can be the hypothesis that an experiment will yield a particular result (i.e. data). Bayes' theorem then takes on the form

$$P(\text{theory}|\text{data}) \propto P(\text{data}|\text{theory}) \cdot P(\text{theory}).$$

Here $P(\text{theory})$ represents the **prior probability** that the theory is true, and $P(\text{data}|\text{theory})$, called the **likelihood**, is the probability, under the assumption of the theory, to observe the data which were actually obtained. The **posterior probability** that the theory is correct after seeing the result of the experiment is then given by $P(\text{theory}|\text{data})$. Here the prior probability for the data $P(\text{data})$ does not appear explicitly, and the equation is expressed as a proportionality. Bayesian statistics provides no fundamental rule for assigning the prior probability to a theory, but once this has been done, it says how one's degree of belief should change in the light of experimental data.

Consider again the probability to have a disease given a positive test result. From the standpoint of someone studying a large number of potential carriers of the disease, the probabilities in this problem can be interpreted as relative frequencies. The prior probability $P(\text{disease})$ is the overall fraction of people who carry the disease, and the posterior probability $P(\text{disease}|+)$ gives the fraction of people who are carriers out of those with a positive test result. A central problem of classical statistics is to estimate the probabilities that are assumed to describe the population as a whole by examining a finite sample of data, i.e. a subsample of the population.

A specific individual, however, may be interested in the subjective probability that he or she has the disease given a positive test result. If no other information is available, one would usually take the prior probability $P(\text{disease})$ to be equal to the overall fraction of carriers, i.e. the same as in the relative frequency interpretation. Here, however, it is taken to mean the degree of belief that one has the disease before taking the test. If other information is available, different prior probabilities could be assigned; this aspect of Bayesian statistics is necessarily subjective, as the name of the probability interpretation implies. Once $P(\text{disease})$ has been assigned, however, Bayes' theorem then tells how the probability to have the disease, i.e. the degree of belief in this hypothesis, changes in light of a positive test result.

The use of subjective probability is discussed further in Sections 6.13, 9.8 and 11.5.3. There exists a vast literature on subjective probability; of particular interest are the books by Jeffreys [Jef48], Savage [Sav72], de Finetti [Fin74] and the paper by Cox [Cox46]. Applications of Bayesian methods are discussed in the books by Lindley [Lin65], O'hagan [Oha94], Lee [Lee89] and Sivia [Siv96].

1.3   Probability density functions

Consider an experiment whose outcome is characterized by a single continuous variable x. The sample space corresponds to the set of possible values that x can assume, and one can ask for the probability of observing a value within an infinitesimal interval $[x, x + dx]$.[2] This is given by the **probability density function** (p.d.f.) $f(x)$:

$$\text{probability to observe } x \text{ in the interval } [x, x + dx] = f(x)dx. \qquad (1.11)$$

In the relative frequency interpretation, $f(x)dx$ gives the fraction of times that x is observed in the interval $[x, x + dx]$ in the limit that the total number of observations is infinitely large. The p.d.f. $f(x)$ is normalized such that the total probability (probability of some outcome) is one,

$$\int_S f(x)dx = 1, \qquad (1.12)$$

where the region of integration S refers to the entire range of x, i.e. to the entire sample space.

Although finite data samples will be dealt with more thoroughly in Chapter 5, it is illustrative here to point out the relationship between a p.d.f. $f(x)$ and a set of n observations of x, $x_1, \ldots, x_n$. A set of such observations can be displayed graphically as a **histogram** as shown in Fig. 1.2. The x axis of the histogram is

---

[2]A possible confusion can arise from the notation used here, since x refers both to the random variable and also to a value that can be assumed by the variable. Many authors use upper case for the random variable, and lower case for the value, i.e. one speaks of X taking on a value in the interval $[x, x + dx]$. This notation is avoided here for simplicity; the distinction between variables and their values should be clear from context.

divided into m subintervals or **bins** of width Δx_i, $i = 1, \ldots, m$, where Δx_i is usually but not necessarily the same for each bin. The number of occurrences n_i of x in subinterval i, i.e. the number of entries in the bin, is given on the vertical axis. The area under the histogram is equal to the total number of entries n multiplied by Δx (or for unequal bin widths, $area = \sum_{i=1}^{m} n_i \cdot \Delta x_i$). Thus the histogram can be normalized to unit area by dividing each n_i by the corresponding bin width Δx_i and by the total number of entries in the histogram n. The p.d.f. $f(x)$ corresponds to a histogram of x normalized to unit area in the limit of zero bin width and an infinitely large total number of entries, as illustrated in Fig. 1.2(d).

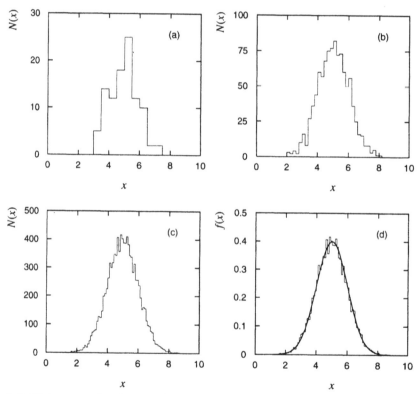

Fig. 1.2 Histograms of various numbers of observations of a random variable x based on the same p.d.f. (a) $n = 100$ observations and a bin width of $\Delta x = 0.5$. (b) $n = 1000$ observations, $\Delta x = 0.2$. (c) $n = 10000$ observations, $\Delta x = 0.1$. (d) The same histogram as in (c), but normalized to unit area. Also shown as a smooth curve is the p.d.f. according to which the observations are distributed. For (a–c), the vertical axis $N(x)$ gives the number of entries in a bin containing x. For (d), the vertical axis is $f(x) = N(x)/(n\Delta x)$.

One can consider cases where the variable x only takes on discrete values x_i, for $i = 1, \ldots, N$, where N can be infinite. The corresponding probabilities can be expressed as

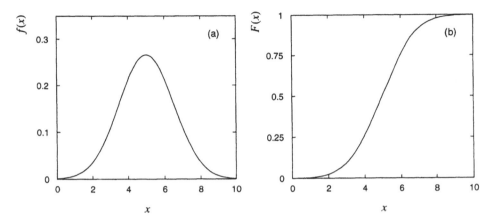

Fig. 1.3 (a) A probability density function $f(x)$. (b) The corresponding cumulative distribution function $F(x)$.

$$\text{probability to observe value } x_i = P(x_i) = f_i, \tag{1.13}$$

where $i = 1, \ldots, N$ and the normalization condition is

$$\sum_{i=1}^{N} f_i = 1. \tag{1.14}$$

Although most of the examples in the following are done with continuous variables, the transformation to the discrete case is a straightforward correspondence between integrals and sums.

The **cumulative distribution** $F(x)$ is related to the p.d.f. $f(x)$ by

$$F(x) = \int_{-\infty}^{x} f(x')dx', \tag{1.15}$$

i.e. $F(x)$ is the probability for the random variable to take on a value less than or equal to x.[3] In fact, $F(x)$ is usually *defined* as the probability to obtain an outcome less than or equal to x, and the p.d.f. $f(x)$ is then defined as $\partial F/\partial x$. For the 'well-behaved' distributions (i.e. $F(x)$ everywhere differentiable) typically encountered in data analysis, the two approaches are equivalent. Figure 1.3 illustrates the relationship between the probability density $f(x)$ and the cumulative distribution $F(x)$.

For a discrete random variable x_i with probabilities $P(x_i)$ the cumulative distribution is defined to be the probability to observe values less than or equal to the value x,

---

[3]Mathematicians call $F(x)$ the 'distribution' function, while physicists often use the word distribution to refer to the probability density function. To avoid confusion we will use the terms cumulative distribution and probability density (or p.d.f.).

$$F(x) = \sum_{x_i \le x} P(x_i). \qquad (1.16)$$

A useful concept related to the cumulative distribution is the so-called **quantile of order** α or α-**point**. The quantile x_α is defined as the value of the random variable x such that $F(x_\alpha) = \alpha$, with $0 \le \alpha \le 1$. That is, the quantile is simply the inverse function of the cumulative distribution,

$$x_\alpha = F^{-1}(\alpha). \qquad (1.17)$$

A commonly used special case is $x_{1/2}$, called the **median** of x. This is often used as a measure of the typical 'location' of the random variable, in the sense that there are equal probabilities for x to be observed greater or less than $x_{1/2}$.

Another commonly used measure of location is the **mode**, which is defined as the value of the random variable at which the p.d.f. is a maximum. A p.d.f. may, of course, have local maxima. By far the most commonly used location parameter is the expectation value, which will be introduced in Section 1.5.

Consider now the case where the result of a measurement is characterized not by one but by several quantities, which may be regarded as a multidimensional random vector. If one is studying people, for example, one might measure for each person their height, weight, age, etc. Suppose a measurement is characterized by two continuous random variables x and y. Let the event A be 'x observed in $[x, x + dx]$ and y observed anywhere', and let B be 'y observed in $[y, y + dy]$ and x observed anywhere', as indicated in Fig. 1.4.

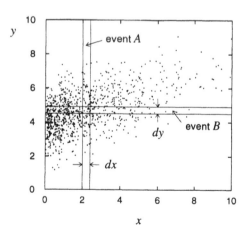

Fig. 1.4 A scatter plot of two random variables x and y based on 1000 observations. The probability for a point to be observed in the square given by the intersection of the two bands (the event $A \cap B$) is given by the joint p.d.f. times the area element, $f(x, y)dxdy$.

The **joint p.d.f.** $f(x, y)$ is defined by

$$
\begin{aligned}
P(A \cap B) \quad &= \quad \text{probability of } x \text{ in } [x, x + dx] \text{ and } y \text{ in } [y, y + dy] \\
&= \quad f(x, y)dxdy. \qquad (1.18)
\end{aligned}
$$

The joint p.d.f. $f(x, y)$ thus corresponds to the density of points on a scatter plot of x and y in the limit of infinitely many points. Since x and y must take on some values, one has the normalization condition

$$\int \int_S f(x, y) dx dy = 1. \tag{1.19}$$

Suppose a joint p.d.f. $f(x, y)$ is known, and one would like to have the p.d.f. for x regardless of the value of y, i.e. corresponding to event A in Fig. 1.4. If one regards the 'event A' column as consisting of squares of area $dx\,dy$, each labeled by an index i, then the probability for A is obtained simply by summing the probabilities corresponding to the individual squares,

$$P(A) = \sum_i f(x, y_i) dy\, dx = f_x(x)\, dx. \tag{1.20}$$

The corresponding probability density, called the **marginal p.d.f.** for x, is then given by the function $f_x(x)$. In the limit of infinitesimal dy, the sum becomes an integral, so that the marginal and joint p.d.f.s are related by

$$f_x(x) = \int_{-\infty}^{\infty} f(x, y) dy. \tag{1.21}$$

Similarly, one obtains the marginal p.d.f. $f_y(y)$ by integrating $f(x, y)$ over x,

$$f_y(y) = \int_{-\infty}^{\infty} f(x, y) dx. \tag{1.22}$$

The marginal p.d.f.s $f_x(x)$ and $f_y(y)$ correspond to the normalized histograms obtained by projecting a scatter plot of x and y onto the respective axes. The relationship between the marginal and joint p.d.f.s is illustrated in Fig. 1.5.

From the definition of conditional probability (1.2), the probability for y to be in $[y, y + dy]$ with any x (event B) given that x is in $[x, x + dx]$ with any y (event A) is

$$P(B|A) = \frac{P(A \cap B)}{P(A)} = \frac{f(x, y) dx dy}{f_x(x) dx}. \tag{1.23}$$

The **conditional p.d.f.** for y given x, $h(y|x)$, is thus defined as

$$h(y|x) = \frac{f(x, y)}{f_x(x)} = \frac{f(x, y)}{\int f(x, y') dy'}. \tag{1.24}$$

This is a p.d.f. of the single random variable y; x is treated as a constant parameter. Starting from $f(x, y)$, one can simply think of holding x constant, and then renormalizing the function such that its area is unity when integrated over y alone.

The conditional p.d.f. $h(y|x)$ corresponds to the normalized histogram of y obtained from the projection onto the y axis of a thin band in x (i.e. with infinitesimal width dx) from an (x, y) scatter plot. This is illustrated in Fig. 1.6 for

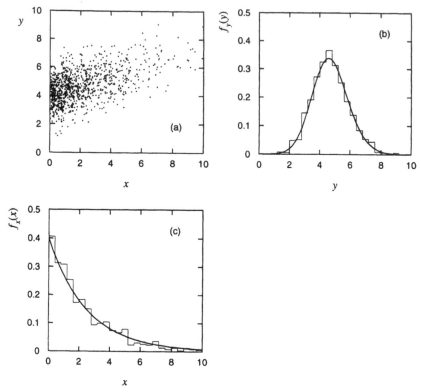

Fig. 1.5 (a) The density of points on the scatter plot is given by the joint p.d.f. $f(x, y)$. (b) Normalized histogram from projecting the points onto the y axis with the corresponding marginal p.d.f. $f_y(y)$. (c) Projection onto the x axis giving $f_x(x)$.

two values of x, leading to two different conditional p.d.f.s, $h(y|x_1)$ and $h(y|x_2)$. Note that $h(y|x_1)$ and $h(y|x_2)$ in Fig. 1.6(b) are both normalized to unit area, as required by the definition of a probability density.

Similarly, the conditional p.d.f. for x given y is

$$g(x|y) = \frac{f(x, y)}{f_y(y)} = \frac{f(x, y)}{\int f(x', y)dx'}. \tag{1.25}$$

Combining equations (1.24) and (1.25) gives the relationship between $g(x|y)$ and $h(y|x)$,

$$g(x|y) = \frac{h(y|x)f_x(x)}{f_y(y)}, \tag{1.26}$$

which is Bayes' theorem for the case of continuous variables (cf. equation (1.6)).

By using $f(x, y) = h(y|x) f_x(x) = g(x|y) f_y(y)$, one can express the marginal p.d.f.s as

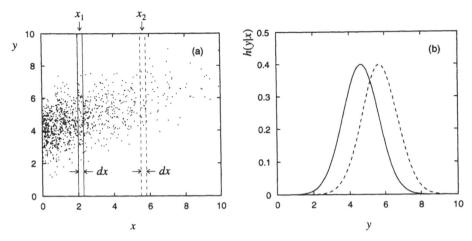

Fig. 1.6 (a) A scatter plot of random variables x and y indicating two infinitesimal bands in x of width dx at x_1 (solid band) and x_2 (dashed band). (b) The conditional p.d.f.s $h(y|x_1)$ and $h(y|x_2)$ corresponding to the projections of the bands onto the y axis.

$$f_x(x) = \int_{-\infty}^{\infty} g(x|y) f_y(y) dy, \tag{1.27}$$

$$f_y(y) = \int_{-\infty}^{\infty} h(y|x) f_x(x) dx. \tag{1.28}$$

These correspond to the law of total probability given by equation (1.7), generalized to the case of continuous random variables.

If 'x in $[x, x + dx]$ with any y' (event A) and 'y in $[y + dy]$ with any x' (event B) are independent, i.e. $P(A \cap B) = P(A) P(B)$, then the corresponding joint p.d.f. for x and y factorizes:

$$f(x, y) = f_x(x) f_y(y). \tag{1.29}$$

From equations (1.24) and (1.25), one sees that for independent random variables x and y the conditional p.d.f. $g(x|y)$ is the same for all y, and similarly $h(y|x)$ does not depend on x. In other words, having knowledge of one of the variables does not change the probabilities for the other. The variables x and y shown in Fig. 1.6, for example, are not independent, as can be seen from the fact that $h(y|x)$ depends on x.

1.4   Functions of random variables

Functions of random variables are themselves random variables. Suppose $a(x)$ is a continuous function of a continuous random variable x, where x is distributed according to the p.d.f. $f(x)$. What is the p.d.f. $g(a)$ that describes the distribution of a? This is determined by requiring that the probability for x to occur between

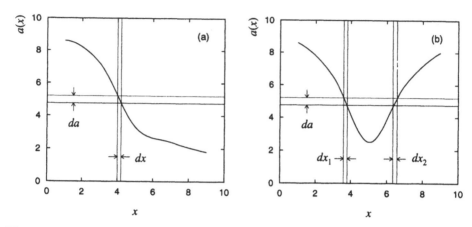

Fig. 1.7   Transformation of variables for (a) a function $q(x)$ with a single-valued inverse $x(a)$ and (b) a function for which the interval da corresponds to two intervals dx_1 and dx_2.

x and $x + dx$ be equal to the probability for a to be between a and $a + da$. That is,

$$g(a')da' = \int_{dS} f(x)dx, \qquad (1.30)$$

where the integral is carried out over the infinitesimal element dS defined by the region in x-space between $a(x) = a'$ and $a(x) = a' + da'$, as shown in Fig. 1.7(a). If the function $a(x)$ can be inverted to obtain $x(a)$, equation (1.30) gives

$$g(a)da = \left| \int_{x(a)}^{x(a+da)} f(x')dx' \right| = \int_{x(a)}^{x(a)+\left|\frac{dx}{da}\right|da} f(x')dx', \qquad (1.31)$$

or

$$g(a) = f(x(a)) \left| \frac{dx}{da} \right|. \qquad (1.32)$$

The absolute value of dx/da ensures that the integral is positive. If the function $a(x)$ does not have a unique inverse, one must include in dS contributions from all regions in x-space between $a(x) = a'$ and $a(x) = a' + da'$, as shown in Fig. 1.7(b).

The p.d.f. $g(a)$ of a function $a(x_1, \ldots, x_n)$ of n random variables $x_1, \ldots, x_n$ with the joint p.d.f. $f(x_1, \ldots, x_n)$ is determined by

$$g(a')da' = \int \cdots \int_{dS} f(x_1, \ldots, x_n)dx_1 \ldots dx_n, \qquad (1.33)$$

where the infinitesimal volume element dS is the region in $x_1, \ldots, x_n$-space between the two (hyper)surfaces defined by $a(x_1, \ldots, x_n) = a'$ and $a(x_1, \ldots, x_n) = a' + da'$.

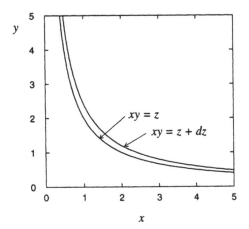

Fig. 1.8 The region of integration dS contained between the two curves $xy = z$ and $xy = z + dz$. Occurrence of (x, y) values between the two curves results in occurrence of z values in the corresponding interval $[z, z + dz]$.

As an example, consider two independent random variables, x and y, distributed according to $g(x)$ and $h(y)$, and suppose we would like to find the p.d.f. of their product $z = xy$. Since x and y are assumed to be independent, their joint p.d.f. is given by $g(x)h(y)$. Equation (1.33) then gives for the p.d.f. of z, $f(z)$,

$$f(z)dz = \int \int_{dS} g(x)h(y)dxdy = \int_{-\infty}^{\infty} g(x)dx \int_{z/|x|}^{(z+dz)/|x|} h(y)dy, \qquad (1.34)$$

where dS is given by the region between $xy = z$ and $xy = z + dz$, as shown in Fig. 1.8. This yields

$$
\begin{aligned}
f(z) &= \int_{-\infty}^{\infty} g(x)h(z/x)\frac{dx}{|x|} \\
&= \int_{-\infty}^{\infty} g(z/y)h(y)\frac{dy}{|y|}, \qquad (1.35)
\end{aligned}
$$

where the second equivalent expression is obtained by reversing the order of integration. Equation (1.35) is often written $f = g \otimes h$, and the function f is called the **Mellin convolution** of g and h.

Similarly, the p.d.f. $f(z)$ of the sum $z = x + y$ is found to be

$$
\begin{aligned}
f(z) &= \int_{-\infty}^{\infty} g(x)h(z - x)dx \\
&= \int_{-\infty}^{\infty} g(z - y)h(y)dy. \qquad (1.36)
\end{aligned}
$$

Equation (1.36) is also often written $f = g \otimes h$, and f is called the **Fourier**

convolution of g and h. In the literature the names Fourier and Mellin are often dropped and one must infer from context what kind of convolution is meant.

Starting from n random variables, $\mathbf{x} = (x_1, \ldots, x_n)$, the following technique can be used to determine the joint p.d.f of n linearly independent functions $a_i(\mathbf{x})$, with $i = 1, \ldots, n$. Assuming the functions $a_1, \ldots, a_n$ can be inverted to give $x_i(a_1, \ldots, a_n)$, $i = 1, \ldots, n$, the joint p.d.f. for the a_i is given by

$$g(a_1, \ldots, a_n) = f(x_1, \ldots, x_n)|J|, \tag{1.37}$$

where $|J|$ is the absolute value of the Jacobian determinant for the transformation,

$$J = \begin{vmatrix} \frac{\partial x_1}{\partial a_1} & \frac{\partial x_1}{\partial a_2} & \cdots & \frac{\partial x_1}{\partial a_n} \\ \frac{\partial x_2}{\partial a_1} & \frac{\partial x_2}{\partial a_2} & \cdots & \frac{\partial x_2}{\partial a_n} \\ \vdots & & & \vdots \\ & & \cdots & \frac{\partial x_n}{\partial a_n} \end{vmatrix}. \tag{1.38}$$

To determine the marginal p.d.f. for one of the functions (say $g_1(a_1)$) the joint p.d.f. $g(a_1, \ldots, a_n)$ must be integrated over the remaining a_i.

In many cases the techniques given above are too difficult to solve analytically. For example, if one is interested in a single function of n random variables, where n is some large and itself possibly variable number, it is rarely practical to come up with $n - 1$ additional functions and then integrate the transformed joint p.d.f. over the unwanted ones. In such cases a numerical solution can usually be found using the Monte Carlo techniques discussed in Chapter 3. If only the mean and variance of a function are needed, the so-called 'error propagation' procedures described in Section 1.6 can be applied.

For certain cases the p.d.f. of a function of random variables can be found using integral transform techniques, specifically, Fourier transforms of the p.d.f.s for sums of random variables and Mellin transforms for products. The basic idea is to take the Mellin or Fourier transform of equation (1.35) or (1.36), respectively. The equation $f = g \otimes h$ is then converted into the product of the transformed density functions, $\tilde{f} = \tilde{g} \cdot \tilde{h}$. The p.d.f. f is obtained by finding the inverse transform of $\tilde{f}$. A complete discussion of these methods is beyond the scope of this book; see e.g. [Spr79]. Some examples of sums of random variables using Fourier transforms (characteristic functions) are given in Chapter 10.

1.5 Expectation values

The **expectation value** $E[x]$ of a random variable x distributed according to the p.d.f. $f(x)$ is defined as

$$E[x] = \int_{-\infty}^{\infty} x f(x) dx = \mu. \tag{1.39}$$

The expectation value of x (also called the **population mean** or simply the mean of x) is often denoted by μ. Note that $E[x]$ is not a function of x, but depends rather on the form of the p.d.f. $f(x)$. If the p.d.f. $f(x)$ is concentrated mostly in one region, then $E[x]$ represents a measure of where values of x are likely to be observed. It can be, however, that $f(x)$ consists of two widely separated peaks, such that $E[x]$ is in the middle where x is seldom (or never) observed.

For a function $a(x)$, the expectation value is

$$E[a] = \int_{-\infty}^{\infty} ag(a)da = \int_{-\infty}^{\infty} a(x)f(x)dx, \qquad (1.40)$$

where $g(a)$ is the p.d.f. of a and $f(x)$ is the p.d.f. of x. The second integral is equivalent; this can be seen by multiplying both sides of equation (1.30) by a and integrating over the entire space.

Some more expectation values of interest are:

$$E[x^n] = \int_{-\infty}^{\infty} x^n f(x)dx = \mu'_n, \qquad (1.41)$$

called the nth algebraic moment of x, for which $\mu = \mu'_1$ is a special case, and

$$E[(x - E[x])^n] = \int_{-\infty}^{\infty} (x - \mu)^n f(x)dx = \mu_n, \qquad (1.42)$$

called the nth central moment of x. In particular, the second central moment,

$$E[(x - E[x])^2] = \int_{-\infty}^{\infty} (x - \mu)^2 f(x)dx = \sigma^2 = V[x], \qquad (1.43)$$

is called the **population variance** (or simply the variance) of x, written σ^2 or $V[x]$. Note that $E[(x - E[x])^2] = E[x^2] - \mu^2$. The variance is a measure of how widely x is spread about its mean value. The square root of the variance σ is called the **standard deviation** of x, which is often useful because it has the same units as x.

For the case of a function a of more than one random variable $\mathbf{x} = (x_1, \ldots, x_n)$, the expectation value is

$$
\begin{aligned}
E[a(\mathbf{x})] &= \int_{-\infty}^{\infty} ag(a)da \\
&= \int_{-\infty}^{\infty} \ldots \int_{-\infty}^{\infty} a(\mathbf{x})f(\mathbf{x})dx_1 \ldots dx_n = \mu_a,
\end{aligned}
\qquad (1.44)
$$

where $g(a)$ is the p.d.f. for a and $f(\mathbf{x})$ is the joint p.d.f. for the x_i. In the following, the notation $\mu_a = E[a]$ will often be used. As in the single-variable case, the two integrals in (1.44) are equivalent, as can be seen by multiplying both sides of equation (1.33) by a and integrating over the entire space. The variance of a is

$$V[a] = E[(a - \mu_a)^2]$$

$$= \int_{-\infty}^{\infty} \cdots \int_{-\infty}^{\infty} (a(\mathbf{x}) - \mu_a)^2 f(\mathbf{x}) dx_1 \ldots dx_n = \sigma_a^2, \qquad (1.45)$$

and is denoted by σ_a^2 or $V[a]$. The **covariance** of two random variables x and y is defined as

$$V_{xy} = E[(x - \mu_x)(y - \mu_y)] = E[xy] - \mu_x \mu_y$$

$$= \int_{-\infty}^{\infty} \int_{-\infty}^{\infty} x\, y\, f(x, y)\, dx\, dy - \mu_x \mu_y, \qquad (1.46)$$

where $\mu_x = E[x]$ and $\mu_y = E[y]$. The covariance matrix V_{xy}, also called the error matrix, is sometimes denoted by $\mathrm{cov}[x, y]$. More generally, for two functions a and b of n random variables $\mathbf{x} = (x_1, \ldots, x_n)$, the covariance $\mathrm{cov}[a, b]$ is given by

$$\mathrm{cov}[a, b] = E[(a - \mu_a)(b - \mu_b)]$$

$$= E[ab] - \mu_a \mu_b$$

$$= \int_{-\infty}^{\infty} \int_{-\infty}^{\infty} a\, b\, g(a, b)\, da\, db - \mu_a \mu_b$$

$$= \int_{-\infty}^{\infty} \cdots \int_{-\infty}^{\infty} a(\mathbf{x})\, b(\mathbf{x})\, f(\mathbf{x}) dx_1 \ldots dx_n - \mu_a \mu_b, \qquad (1.47)$$

where $g(a, b)$ is the joint p.d.f. for a and b and $f(\mathbf{x})$ is the joint p.d.f. for the x_i. As in equation (1.44), the two integral expressions for V_{ab} are equivalent. Note that by construction the covariance matrix V_{ab} is symmetric in a and b and that the diagonal elements $V_{aa} = \sigma_a^2$ (i.e. the variances) are positive.

In order to give a dimensionless measure of the level of correlation between two random variables x and y, one often uses the **correlation coefficient**, defined by

$$\rho_{xy} = \frac{V_{xy}}{\sigma_x \sigma_y}. \qquad (1.48)$$

One can show (see e.g. [Fro79, Bra92]) that the correlation coefficient lies in the range $-1 \leq \rho_{xy} \leq 1$.

One can roughly understand the covariance of two random variables x and y in the following way. V_{xy} is the expectation value of $(x - \mu_x)(y - \mu_y)$, the product of the deviations of x and y from their means, μ_x and μ_y. Suppose that

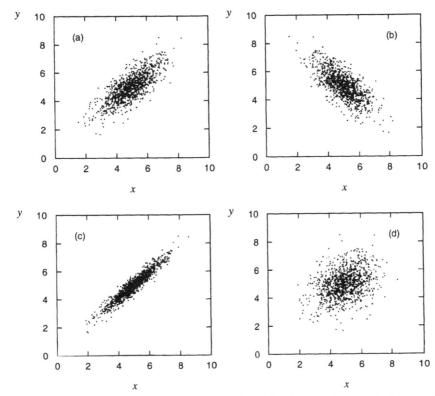

Fig. 1.9 Scatter plots of random variables x and y with (a) a positive correlation, $\rho = 0.75$, (b) a negative correlation, $\rho = -0.75$, (c) $\rho = 0.95$, and (d) $\rho = 0.25$. For all four cases the standard deviations of x and y are $\sigma_x = \sigma_y = 1$.

having x greater than μ_x enhances the probability to find y greater than μ_y, and x less than μ_x gives an enhanced probability to have y less than μ_y. Then V_{xy} is greater than zero, and the variables are said to be positively correlated. Such a situation is illustrated in Figs 1.9 (a), (c) and (d), for which the correlation coefficients ρ_{xy} are 0.75, 0.95 and 0.25, respectively. Similarly, $V_{xy} < 0$ is called a negative correlation: having $x > \mu_x$ increases the probability to observe $y < \mu_y$. An example is shown in Fig. 1.9(b), for which $\rho_{xy} = -0.75$.

From equations (1.29) and (1.44), it follows that for independent random variables x and y,

$$E[xy] = E[x]E[y] = \mu_x \mu_y, \tag{1.49}$$

(and hence by equation (1.46), $V_{xy} = 0$) although the converse is not necessarily true. Figure 1.10, for example, shows a two-dimensional scatter plot of a p.d.f. for which $V_{xy} = 0$, but where x and y are not independent. That is, $f(x, y)$ does not factorize according to equation (1.29), and hence knowledge of one of the

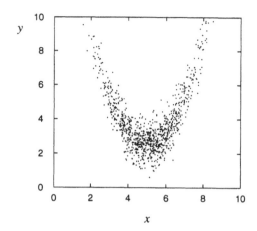

y

Fig. **1.10** Scatter plot of random variables x and y which are not independent (i.e. $f(x,y) \neq f_x(x)f_y(y)$) but for which $V_{xy} = 0$ because of the particular symmetry of the distribution.

x

variables affects the conditional p.d.f. of the other. The covariance V_{xy} vanishes, however, because $f(x,y)$ is symmetric in x about the mean μ_x.

1.6 Error propagation

Suppose one has a set of n random variables $\mathbf{x} = (x_1, \ldots, x_n)$ distributed according to some joint p.d.f. $f(\mathbf{x})$. Suppose that the p.d.f. is not completely known, but the mean values of the x_i, $\boldsymbol{\mu} = (\mu_1, \ldots, \mu_n)$, and the covariance matrix, V_{ij}, are known or have at least been estimated. (Methods for doing this are described in Chapter 5.)

Now consider a function of the n variables $y(\mathbf{x})$. To determine the p.d.f. for y, one must in principle follow a procedure such as those described in Section 1.4 (e.g. equations (1.33) or (1.37)). We have assumed, however, that $f(\mathbf{x})$ is not completely known, only the means $\boldsymbol{\mu}$ and the covariance matrix V_{ij}, so this is not possible. One can, however, approximate the expectation value of y and the variance $V[y]$ by first expanding the function $y(\mathbf{x})$ to first order about the mean values of the x_i,

$$y(\mathbf{x}) \approx y(\boldsymbol{\mu}) + \sum_{i=1}^{n} \left[\frac{\partial y}{\partial x_i} \right]_{\mathbf{x}=\boldsymbol{\mu}} (x_i - \mu_i). \tag{1.50}$$

The expectation value of y is to first order

$$E[y(\mathbf{x})] \approx y(\boldsymbol{\mu}), \tag{1.51}$$

since $E[x_i - \mu_i] = 0$. The expectation value of y^2 is

$$E[y^2(\mathbf{x})] \approx y^2(\boldsymbol{\mu}) + 2y(\boldsymbol{\mu}) \cdot \sum_{i=1}^{n} \left[\frac{\partial y}{\partial x_i}\right]_{\mathbf{x}=\boldsymbol{\mu}} E[x_i - \mu_i]$$

$$+ \quad E\left[\left(\sum_{i=1}^{n} \left[\frac{\partial y}{\partial x_i}\right]_{\mathbf{x}=\boldsymbol{\mu}} (x_i - \mu_i)\right)\left(\sum_{j=1}^{n} \left[\frac{\partial y}{\partial x_j}\right]_{\mathbf{x}=\boldsymbol{\mu}} (x_j - \mu_j)\right)\right]$$

$$= \quad y^2(\boldsymbol{\mu}) + \sum_{i,j=1}^{n} \left[\frac{\partial y}{\partial x_i}\frac{\partial y}{\partial x_j}\right]_{\mathbf{x}=\boldsymbol{\mu}} V_{ij}, \tag{1.52}$$

so that the variance $\sigma_y^2 = E[y^2] - (E[y])^2$ is given by

$$\sigma_y^2 \approx \sum_{i,j=1}^{n} \left[\frac{\partial y}{\partial x_i}\frac{\partial y}{\partial x_j}\right]_{\mathbf{x}=\boldsymbol{\mu}} V_{ij}. \tag{1.53}$$

Similarly, one obtains for a set of m functions $y_1(\mathbf{x}), \ldots, y_m(\mathbf{x})$ the covariance matrix

$$U_{kl} = \text{cov}[y_k, y_l] \approx \sum_{i,j=1}^{n} \left[\frac{\partial y_k}{\partial x_i}\frac{\partial y_l}{\partial x_j}\right]_{\mathbf{x}=\boldsymbol{\mu}} V_{ij}. \tag{1.54}$$

This can be expressed in matrix notation as

$$U = A\,V\,A^T, \tag{1.55}$$

where the matrix of derivatives A is

$$A_{ij} = \left[\frac{\partial y_i}{\partial x_j}\right]_{\mathbf{x}=\boldsymbol{\mu}} \tag{1.56}$$

and A^T is the transpose of A. Equations (1.53)–(1.56) form the basis of **error propagation** (i.e. the variances, which are used as measures of statistical uncertainties, are propagated from the x_i to the functions y_1, y_2, etc.). (The term 'error' will often be used to refer to the uncertainty of a measurement, which in most cases is given by the standard deviation of the corresponding random variable.)

For the case where the x_i are not correlated, i.e. $V_{ii} = \sigma_i^2$ and $V_{ij} = 0$ for $i \neq j$, equations (1.53) and (1.54) become

$$\sigma_y^2 \approx \sum_{i=1}^{n} \left[\frac{\partial y}{\partial x_i}\right]_{\mathbf{x}=\boldsymbol{\mu}}^{2} \sigma_i^2 \tag{1.57}$$

and

$$U_{kl} \approx \sum_{i=1}^{n} \left[\frac{\partial y_k}{\partial x_i} \frac{\partial y_l}{\partial x_i} \right]_{\mathbf{x}=\boldsymbol{\mu}} \sigma_i^2. \tag{1.58}$$

Equation (1.53) leads to the following special cases. If $y = x_1 + x_2$, the variance of y is then

$$\sigma_y^2 = \sigma_1^2 + \sigma_2^2 + 2V_{12}. \tag{1.59}$$

For the product $y = x_1 x_2$ one obtains

$$\frac{\sigma_y^2}{y^2} = \frac{\sigma_1^2}{x_1^2} + \frac{\sigma_2^2}{x_2^2} + 2\frac{V_{12}}{x_1 x_2}. \tag{1.60}$$

If the variables x_1 and x_2 are not correlated ($V_{12} = 0$), the relations above state that errors (i.e. standard deviations) add quadratically for the sum $y = x_1 + x_2$, and that the *relative* errors add quadratically for the product $y = x_1 x_2$.

In deriving the error propagation formulas we have assumed that the means and covariances of the original set of variables $x_1, \ldots, x_n$ are known (or at least estimated) and that the desired functions of these variables can be approximated by the first-order Taylor expansion around the means $\mu_1, \ldots, \mu_n$. The latter assumption is of course only exact for a linear function. The approximation breaks down if the function $y(\mathbf{x})$ (or functions $\mathbf{y}(\mathbf{x})$) are significantly nonlinear in a region around the means $\boldsymbol{\mu}$ of a size comparable to the standard deviations of the x_i, $\sigma_1, \ldots, \sigma_n$. Care must be taken, for example, with functions like $y(x) = 1/x$ when $E[x] = \mu$ is comparable to or smaller than the standard deviation of x. Such situations can be better treated with the Monte Carlo techniques described in Chapter 3, or using confidence intervals as described in Section 9.2.

1.7 Orthogonal transformation of random variables

Suppose one has a set of n random variables $x_1, \ldots, x_n$ and their covariance matrix $V_{ij} = \text{cov}[x_i, x_j]$, for which the off-diagonal elements are not necessarily zero. Often it can be useful to define n new variables $y_1, \ldots, y_n$ that are not correlated, i.e. for which the new covariance matrix $U_{ij} = \text{cov}[y_i, y_j]$ is diagonal. We will show that this is always possible with a linear transformation,

$$y_i = \sum_{j=1}^{n} A_{ij} x_j. \tag{1.61}$$

Assuming such a transformation, the covariance matrix for the new variables is

$$U_{ij} = \text{cov}[y_i, y_j] = \text{cov}\left[\sum_{k=1}^{n} A_{ik}x_k, \sum_{l=1}^{n} A_{jl}x_l\right]$$

$$= \sum_{k,l=1}^{n} A_{ik}A_{jl}\,\text{cov}[x_k, x_l]$$

$$= \sum_{k,l=1}^{n} A_{ik}V_{kl}A_{lj}^{T}. \tag{1.62}$$

This is simply a special case of the error propagation formula (1.54); here it is exact, since the function (1.61) is linear.

The problem thus consists of finding a matrix A such that $U = AVA^T$ is diagonal. This is simply the diagonalization of a real, symmetric matrix, a well-known problem of linear algebra (cf. [Arf95]). The solution can be found by first determining the eigenvectors $\mathbf{r}^i$, $i = 1, \ldots, n$, of the covariance matrix V. That is, one must solve the equation

$$V\mathbf{r}^i = \lambda_i \mathbf{r}^i, \tag{1.63}$$

where in the matrix equations the vector $\mathbf{r}$ should be understood as a column vector. The eigenvectors $\mathbf{r}^i$ are only determined up to a multiplicative factor, which can be chosen such that they all have unit length. Furthermore, one can easily show that since the covariance matrix is symmetric, the eigenvectors are orthogonal, i.e.

$$\mathbf{r}^i \cdot \mathbf{r}^j = \sum_{k=1}^{n} r_k^i r_k^j = \delta_{ij}. \tag{1.64}$$

If two or more of the eigenvalues $\lambda_i, \lambda_j, \ldots$ are equal, then the directions of the corresponding eigenvectors $\mathbf{r}^i, \mathbf{r}^j, \ldots$ are not uniquely determined, but can nevertheless be chosen such that the eigenvectors are orthogonal.

The n rows of the transformation matrix A are then given by the n eigenvectors $\mathbf{r}^i$ (in any order), i.e. $A_{ij} = r_j^i$, and the transpose matrix thus has the eigenvectors as its columns, $A_{ij}^T = r_i^j$. That this matrix has the desired property can be shown explicitly by substituting it into equation (1.62),

$$U_{ij} = \sum_{k,l=1}^{n} A_{ik}V_{kl}A_{lj}^{T} = \sum_{k,l=1}^{n} r_k^i V_{kl} r_l^j$$

$$= \sum_{k=1}^{n} r_k^i \lambda_j r_k^j$$

$$= \lambda_j \mathbf{r}^i \cdot \mathbf{r}^j$$

$$= \lambda_j \delta_{ij}. \tag{1.65}$$

Thus the variances of the transformed variables $y_1, \ldots, y_n$ are given by the eigenvalues of the original covariance matrix V, and all off-diagonal elements of U are zero. Since the eigenvectors are orthonormal (equation (1.64)), one has the property

$$\sum_{j=1}^{n} A_{ij} A_{jk}^T = \sum_{j=1}^{n} r_j^i r_j^k = \mathbf{r}^i \cdot \mathbf{r}^k = \delta_{ik}, \tag{1.66}$$

or as a matrix equation $AA^T = 1$, and hence $A^T = A^{-1}$. Such a transformation is said to be orthogonal, i.e. it corresponds to a rotation of the vector $\mathbf{x}$ into $\mathbf{y}$ such that the norm remains constant, since $|\mathbf{y}|^2 = \mathbf{y}^T \mathbf{y} = \mathbf{x}^T A^T A \mathbf{x} = |\mathbf{x}|^2$.

In order to find the eigenvectors of V, the standard techniques of linear algebra can used (see e.g. [Arf95]). For more than three variables, the problem becomes impractical to solve analytically, and numerical techniques such as the **singular value decomposition** are necessary (see e.g. [Bra92, Pre92]).

In two dimensions, for example, the covariance matrix for the variables $\mathbf{x} = (x_1, x_2)$ can be expressed as

$$V = \begin{pmatrix} \sigma_1^2 & \rho\sigma_1\sigma_2 \\ \rho\sigma_1\sigma_2 & \sigma_2^2 \end{pmatrix}. \tag{1.67}$$

The eigenvalue equation $(V - I\lambda)\mathbf{r} = 0$ (where I is the 2×2 unit matrix) is solved by requiring that the determinant of the matrix of coefficients be equal to zero,

$$\det(V - I\lambda) = 0. \tag{1.68}$$

The two eigenvalues $\lambda_{\pm}$ are found to be

$$\lambda_{\pm} = \tfrac{1}{2}\left[\sigma_1^2 + \sigma_2^2 \pm \sqrt{(\sigma_1^2 + \sigma_2^2)^2 - 4(1 - \rho^2)\sigma_1^2\sigma_2^2}\right]. \tag{1.69}$$

The two orthonormal eigenvectors $\mathbf{r}_{\pm}$ can be parametrized by an angle θ,

$$\mathbf{r}_+ = \begin{pmatrix} \cos\theta \\ \sin\theta \end{pmatrix} \qquad \mathbf{r}_- = \begin{pmatrix} -\sin\theta \\ \cos\theta \end{pmatrix}. \tag{1.70}$$

Substituting the eigenvalues (1.69) back into the eigenvalue equation determines the angle θ,

$$\theta = \tfrac{1}{2}\tan^{-1}\left(\frac{2\rho\sigma_1\sigma_2}{\sigma_1^2 - \sigma_2^2}\right). \tag{1.71}$$

The rows of the desired transformation matrix are thus given by the two eigenvectors,

$$A = \begin{pmatrix} \cos\theta & \sin\theta \\ -\sin\theta & \cos\theta \end{pmatrix} . \qquad (1.72)$$

This corresponds to a rotation of the vector (x_1, x_2) by an angle θ. An example is shown in Fig. 1.11 where the original two variables have $\sigma_1 = 1.5$, $\sigma_2 = 1.0$, and a correlation coefficient of $\rho = 0.7$.

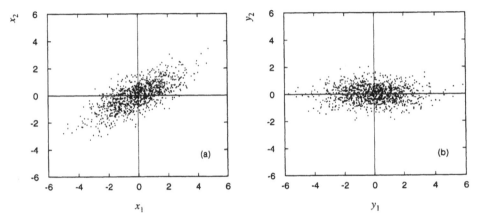

Fig. 1.11 Scatter plot of (a) two correlated random variables (x_1, x_2) and (b) the transformed variables (y_1, y_2) for which the covariance matrix is diagonal.

Although uncorrelated variables are often easier to deal with, the transformed variables may not have as direct an interpretation as the original ones. Examples where this procedure could be used will arise in Chapters 6 through 8 on parameter estimation, where the estimators for a set of parameters will often be correlated.

2

Examples of probability functions

In this chapter a number of commonly used probability distributions and density functions are presented. Properties such as mean and variance are given, mostly without proof; the moments can be found by using characteristic functions introduced in Chapter 10. Additional p.d.f.s can be found in [Fro79] Chapter 4, [Ead71] Chapter 4, [Bra92] Chapter 5.

2.1 Binomial and multinomial distributions

Consider a series of N independent trials or observations, each having two possible outcomes, here called 'success' and 'failure', where the probability for success is some constant value, p. The set of trials can be regarded as a single measurement and is characterized by a discrete random variable n, defined to be the total number of successes. That is, the sample space is defined to be the set of possible values of n successes given N observations. If one were to repeat the entire experiment many times with N trials each time, the resulting values of n would occur with relative frequencies given by the so-called **binomial distribution**.

The form of the binomial distribution can be derived in the following way. We have assumed that the probability of success in a single observation is p and the probability of failure is $1 - p$. Since the individual trials are assumed to be independent, the probability for a series of successes and failures in a particular order is equal to the product of the individual probabilities. For example, the probability in five trials to have success, success, failure, success, failure in that order is $p \cdot p \cdot (1 - p) \cdot p \cdot (1 - p) = p^3(1 - p)^2$. In general the probability for a particular sequence of n successes and $N - n$ failures is $p^n(1 - p)^{N-n}$. We are not interested in the order, however, only in the final number of successes n. The number of sequences having n successes in N events is

$$\frac{N!}{n!(N - n)!},$$
(2.1)

so the total probability to have n successes in N events is

$$f(n; N, p) = \frac{N!}{n!(N - n)!} p^n (1 - p)^{N-n},$$
(2.2)

for $n = 0, \ldots, N$. Note that $f(n; N, p)$ is itself a probability, not a probability density. The notation used is that the random variable (or variables) are listed as arguments of the probability function (or p.d.f.) to the left of the semicolon, and any parameters (in this case N and p) are listed to the right. The expectation value of n is

$$E[n] = \sum_{n=0}^{\infty} n \, \frac{N!}{n!(N-n)!} \, p^n \, (1-p)^{N-n} = Np, \qquad (2.3)$$

and variance is

$$\begin{aligned} V[n] &= E[n^2] - (E[n])^2 \\ &= Np(1-p). \end{aligned} \qquad (2.4)$$

These can be computed by using the characteristic function of the binomial distribution, cf. Chapter 10.

Recall that expectation values are not functions of the random variable, but they depend on the parameters of the probability function, in this case p and N. The binomial probability distribution is shown in Fig. 2.1 and Fig. 2.2 for various values of p and N.

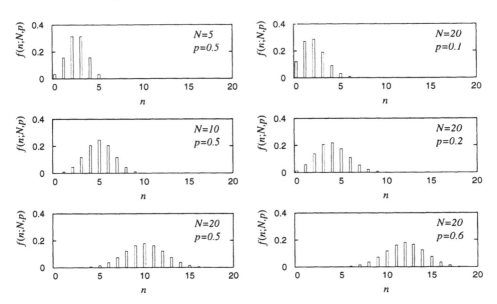

Fig. 2.1 The binomial distribution for $p = 0.5$ and various values of N.

Fig. 2.2 The binomial distribution for $N = 20$ and various values of p.

The **multinomial distribution** is the generalization of the binomial distribution to the case where there are not only two outcomes ('success' and 'failure') but rather m different possible outcomes. For a particular trial the probability of outcome i is p_i, and since one of the outcomes must be realized, one has the constraint $\sum_{i=1}^{m} p_i = 1$.

Now consider a measurement consisting of N trials, each of which yields one of the possible m outcomes. The probability for a particular sequence of outcomes, e.g. i on the first trial, j on the second, and so on, in a particular order, is the product of the N corresponding probabilities, $p_i p_j \ldots p_k$. The number of such sequences that will lead to n_1 outcomes of type 1, n_2 outcomes of type 2, etc., is

$$\frac{N!}{n_1! n_2! \ldots n_m!}. \tag{2.5}$$

If we are not interested in the order of the outcomes, but only in the total numbers of each type, then the joint probability for n_1 outcomes of type 1, n_2 of type 2, etc.. is given by the multinomial distribution,

$$f(n_1, \ldots, n_m; N, p_1, \ldots, p_m) = \frac{N!}{n_1! n_2! \ldots n_m!} p_1^{n_1} p_2^{n_2} \cdots p_m^{n_m}. \tag{2.6}$$

Suppose one breaks the m possible outcomes into two categories: outcome i ('success') and not outcome i ('failure'). Since this is the same as the binomial process presented above, the number of occurrences of outcome i, n_i, must be binomially distributed. This is of course true for all i. From equations (2.3) and (2.4) one has that the expectation value of n_i is $E[n_i] = N p_i$ and the variance is $V[n_i] = N p_i (1 - p_i)$.

Consider now the three possible outcomes: i, j and everything else. The probability to have n_i outcomes of type i, n_j of type j and $N - n_i - n_j$ of everything else is

$$f(n_i, n_j; N, p_i, p_j) = \frac{N!}{n_i! n_j! (N - n_i - n_j)!} p_i^{n_i} p_j^{n_j} (1 - p_i - p_j)^{N - n_i - n_j}, \tag{2.7}$$

so that the covariance $V_{ij} = \text{cov}[n_i, n_j]$ is

$$\begin{aligned} V_{ij} &= E[(n_i - E[n_i])(n_j - E[n_j])] \\ &= -N p_i p_j \end{aligned} \tag{2.8}$$

for $i \neq j$, otherwise $V_{ii} = \sigma_i^2 = N p_i (1 - p_i)$.

An example of the multinomial distribution is the probability to obtain a particular result for a histogram constructed from N independent observations of a random variable, i.e. n_1 entries in bin 1, n_2 entries in bin 2, etc., with m bins and N total entries. Note from equation (2.8) that the numbers of entries in any two bins are negatively correlated. That is, if in N trials bin i contains a larger than average number of entries ($n_i > N p_i$) then the probability is increased that a different bin j will contain a smaller than average number.

2.2 Poisson distribution

Consider the binomial distribution of Section 2.1 in the limit that N becomes very large, p becomes very small, but the product Np (i.e. the expectation value of the number of successes) remains equal to some finite value ν. It can be shown that equation (2.2) leads in this limit to (see Section 10.2)

$$f(n;\nu) = \frac{\nu^n}{n!} e^{-\nu}, \tag{2.9}$$

which is called the **Poisson distribution** for the integer random variable n, where $n = 0, 1, \ldots, \infty$. The p.d.f. has one parameter, ν. Figure 2.3 shows the Poisson distribution for $\nu = 2, 5, 10$.

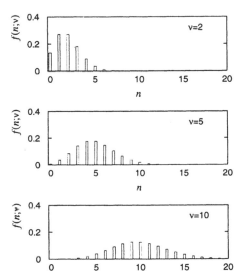

Fig. 2.3 The Poisson probability distribution for various values of the parameter ν.

The expectation value of the Poisson random variable n is

$$E[n] = \sum_{n=0}^{\infty} n \frac{\nu^n}{n!} e^{-\nu} = \nu, \tag{2.10}$$

and the variance is given by

$$V[n] = \sum_{n=0}^{\infty} (n - \nu)^2 \frac{\nu^n}{n!} e^{-\nu} = \nu. \tag{2.11}$$

Although a Poisson variable is discrete, it can be treated as a continuous variable x as long as this is integrated over a range Δx which is large compared to unity. We will show in Chapter 10 that for large mean value ν, a Poisson variable can be treated as a continuous variable following a Gaussian distribution, cf. Section 2.5.

An example of a Poisson random variable is the number of decays of a certain amount of radioactive material in a fixed time period, in the limit that the total number of possible decays (i.e. the total number of radioactive atoms) is very large and the probability for an individual decay within the time period is very small. Another example is the number of events of a certain type observed in a particle scattering experiment with a given integrated luminosity L. The expectation value of the number of events is

$$\nu = \sigma L \varepsilon, \tag{2.12}$$

where σ is the cross section for the event and ε is the efficiency, i.e. the probability for the event to be observed in the detector.

2.3   Uniform distribution

The **uniform** p.d.f. for the continuous variable x $(-\infty < x < \infty)$ is defined by

$$f(x; \alpha, \beta) = \begin{cases} \frac{1}{\beta - \alpha} & \alpha \le x \le \beta \\ 0 & \text{otherwise,} \end{cases} \tag{2.13}$$

i.e. x is equally likely to be found anywhere between α and β. The mean and variance of x are given by

$$E[x] = \int_\alpha^\beta \frac{x}{\beta - \alpha} \, dx = \tfrac{1}{2}(\alpha + \beta), \tag{2.14}$$

$$V[x] = \int_\alpha^\beta [x - \tfrac{1}{2}(\alpha + \beta)]^2 \frac{1}{\beta - \alpha} \, dx = \tfrac{1}{12}(\beta - \alpha)^2. \tag{2.15}$$

An important feature of the uniform distribution is that any continuous random variable x with p.d.f. $f(x)$ and cumulative distribution $F(x)$ can easily be transformed to a new variable y which is uniformly distributed between zero and one. The transformed variable y is simply given by

$$y = F(x), \tag{2.16}$$

i.e. it is the cumulative distribution function of the original variable x. For any cumulative distribution $y = F(x)$ one has

$$\frac{dy}{dx} = \frac{d}{dx} \int_{-\infty}^x f(x') \, dx' = f(x), \tag{2.17}$$

and hence from equation (1.32) one finds the p.d.f. of y to be

$$g(y) = f(x) \left| \frac{dx}{dy} \right| = f(x) \left| \frac{dy}{dx} \right|^{-1} = 1, \quad (0 \le y \le 1). \tag{2.18}$$

This property of the uniform distribution will be used in Chapter 3 in connection with Monte Carlo techniques.

An example of a uniformly distributed variable is the energy of a photon (γ) from the decay of a neutral pion, $\pi^0 \rightarrow \gamma\gamma$. The laboratory-frame energy E_γ of either photon is uniformly distributed between $E_{\min} = \frac{1}{2}E_\pi(1 - \beta)$ and $E_{\max} = \frac{1}{2}E_\pi(1 + \beta)$, where E_π is the energy of the pion and $\beta = v/c$ is the pion's velocity divided by the velocity of light.

2.4 Exponential distribution

The **exponential** probability density of the continuous variable x ($0 \leq x < \infty$) is defined by

$$f(x;\xi) = \frac{1}{\xi}e^{-x/\xi}. \tag{2.19}$$

The p.d.f. is characterized by a single parameter ξ. The expectation value of x is

$$E[x] = \frac{1}{\xi}\int_0^\infty x e^{-x/\xi}dx = \xi, \tag{2.20}$$

and the variance of x is given by

$$V[x] = \frac{1}{\xi}\int_0^\infty (x - \xi)^2 e^{-x/\xi}dx = \xi^2. \tag{2.21}$$

An example of an exponential random variable is the decay time of an unstable particle measured in its rest frame. The parameter ξ then corresponds to the mean lifetime, usually denoted by τ. The exponential distribution is shown in Fig. 2.4 for different values of ξ.

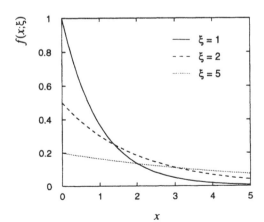

Fig. 2.4 The exponential probability density for various values of the parameter ξ.

2.5   Gaussian distribution

The **Gaussian** (or **normal**) p.d.f. of the continuous random variable x (with $-\infty < x < \infty$) is defined by

$$f(x; \mu, \sigma^2) = \frac{1}{\sqrt{2\pi\sigma^2}} \exp\left(\frac{-(x - \mu)^2}{2\sigma^2}\right),\qquad(2.22)$$

which has two parameters, μ and σ^2. The names of the parameters are clearly motivated by the values of the mean and variance of x. These are found to be

$$E[x] = \int_{-\infty}^{\infty} x \frac{1}{\sqrt{2\pi\sigma^2}} \exp\left(\frac{-(x - \mu)^2}{2\sigma^2}\right) dx = \mu,\qquad(2.23)$$

$$V[x] = \int_{-\infty}^{\infty} (x - \mu)^2 \frac{1}{\sqrt{2\pi\sigma^2}} \exp\left(\frac{-(x - \mu)^2}{2\sigma^2}\right) dx = \sigma^2.\qquad(2.24)$$

Recall that μ and σ^2 are often used to denote the mean and variance of any p.d.f. as defined by equations (1.39) and (1.43), not only those of a Gaussian. Note also that one may equivalently regard either σ or σ^2 as the parameter. The Gaussian p.d.f. is shown in Fig. 2.5 for different combinations of the parameters μ and σ.

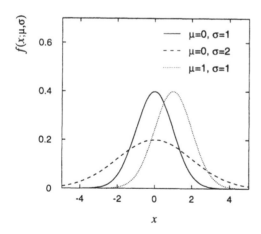

Fig. 2.5 The Gaussian probability density for various values of the parameters μ and σ.

A special case of the Gaussian p.d.f. is sufficiently important to merit its own notation. Using $\mu = 0$ and $\sigma = 1$, one defines the **standard Gaussian** p.d.f. $\varphi(x)$ as

$$\varphi(x) = \frac{1}{\sqrt{2\pi}} \exp(-x^2/2),\qquad(2.25)$$

with the corresponding cumulative distribution $\Phi(x)$,

$$\Phi(x) = \int_{-\infty}^{x} \varphi(x')dx'. \qquad (2.26)$$

One can easily show that if y is distributed according to a Gaussian p.d.f. with mean μ and variance σ^2, then the variable

$$x = \frac{y - \mu}{\sigma} \qquad (2.27)$$

is distributed according to the standard Gaussian $\varphi(x)$, and the cumulative distributions are related by $F(y) = \Phi(x)$. The cumulative distribution $\Phi(x)$ cannot be expressed analytically and must be evaluated numerically. Values of $\Phi(x)$ as well as the quantiles $x_\alpha = \Phi^{-1}(\alpha)$ are tabulated in many reference books (e.g. [Bra92, Fro79, Dud88]) and are also available from computer program libraries, e.g. the routines FREQ and GAUSIN in [CER97].

The importance of the Gaussian distribution stems from the **central limit theorem**. The theorem states that the sum of n independent continuous random variables x_i with means μ_i and variances σ_i^2 becomes a Gaussian random variable with mean $\mu = \sum_{i=1}^{n} \mu_i$ and variance $\sigma^2 = \sum_{i=1}^{n} \sigma_i^2$ in the limit that n approaches infinity. This holds (under fairly general conditions) regardless of the form of the individual p.d.f.s of the x_i. This is the formal justification for treating measurement errors as Gaussian random variables, and holds to the extent that the total error is the sum of a large number of small contributions. The theorem can be proven using characteristic functions as described in Section 10.3.

The N-dimensional generalization of the Gaussian distribution is defined by

$$f(\mathbf{x}; \boldsymbol{\mu}, V) = \frac{1}{(2\pi)^{N/2}|V|^{1/2}} \exp\left[-\frac{1}{2}(\mathbf{x} - \boldsymbol{\mu})^T V^{-1}(\mathbf{x} - \boldsymbol{\mu})\right], \qquad (2.28)$$

where $\mathbf{x}$ and $\boldsymbol{\mu}$ are column vectors containing $x_1, \ldots, x_N$ and $\mu_1, \ldots, \mu_N$, $\mathbf{x}^T$ and $\boldsymbol{\mu}^T$ are the corresponding row vectors, and $|V|$ is the determinant of a symmetric $N \times N$ matrix V, thus containing $N(N+1)/2$ free parameters. For now regard V as a label for the parameters of the Gaussian, although as with the one-dimensional case, the notation is motivated by what one obtains for the covariance matrix. The expectation values and (co)variances can be computed to be

$$\begin{aligned} E[x_i] &= \mu_i \\ V[x_i] &= V_{ii} \\ \mathrm{cov}[x_i, x_j] &= V_{ij}. \end{aligned} \qquad (2.29)$$

For two dimensions the p.d.f. becomes

$$f(x_1, x_2; \mu_1, \mu_2, \sigma_1, \sigma_2, \rho) = \frac{1}{2\pi\sigma_1\sigma_2\sqrt{1-\rho^2}}$$
$$\times \exp\left\{-\frac{1}{2(1-\rho^2)}\left[\left(\frac{x_1-\mu_1}{\sigma_1}\right)^2 + \left(\frac{x_2-\mu_2}{\sigma_2}\right)^2 - 2\rho\left(\frac{x_1-\mu_1}{\sigma_1}\right)\left(\frac{x_2-\mu_2}{\sigma_2}\right)\right]\right\},$$

(2.30)

where $\rho = \text{cov}[x_1, x_2]/(\sigma_1\sigma_2)$ is the correlation coefficient.

2.6 Log-normal distribution

If a continuous variable y is Gaussian with mean μ and variance σ^2, then $x = e^y$ follows the **log-normal** distribution. This is given by

$$f(x; \mu, \sigma^2) = \frac{1}{\sqrt{2\pi\sigma^2}}\frac{1}{x}\exp\left(\frac{-(\log x - \mu)^2}{2\sigma^2}\right).$$

(2.31)

The expectation value and variance are given in terms of the two parameters μ and σ^2 as

$$E[x] = \exp(\mu + \tfrac{1}{2}\sigma^2),$$

(2.32)

$$V[x] = \exp(2\mu + \sigma^2)[\exp(\sigma^2) - 1].$$

(2.33)

As in the case of the Gaussian p.d.f., one may consider either σ^2 or σ as the parameter. Note that here, however, μ and σ^2 are not the mean and variance of x, but rather the parameters of the corresponding Gaussian distribution for $\log x$. The log-normal p.d.f. is shown in Fig. 2.6 for several combinations of μ and σ.

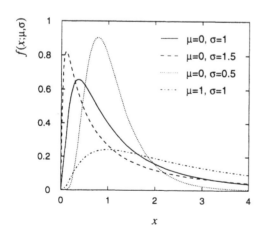

Fig. 2.6 The log-normal probability density for several values of the parameters μ and σ.

Recall from the central limit theorem, introduced in the previous section, that if a random variable y is the sum of a large number of small contributions, then it will be distributed according to a Gaussian p.d.f. From this it follows that if a variable x is given by the product of many factors then it will follow a log-normal distribution. It can thus be used to model random errors which change a result by a multiplicative factor.

2.7 Chi-square distribution

The χ^2 (**chi-square**) distribution of the continuous variable z ($0 \leq z < \infty$) is defined by

$$f(z;n) = \frac{1}{2^{n/2}\Gamma(n/2)} z^{n/2-1} e^{-z/2}, \ n = 1, 2, \ldots, \tag{2.34}$$

where the parameter n is called the **number of degrees of freedom**, and the gamma function $\Gamma(x)$ is defined by

$$\Gamma(x) = \int_0^\infty e^{-t} t^{x-1} \, dt. \tag{2.35}$$

For the purposes of computing the χ^2 distribution, one only needs to know that $\Gamma(n) = (n-1)!$ for integer n, $\Gamma(x+1) = x\Gamma(x)$, and $\Gamma(1/2) = \sqrt{\pi}$. The mean and variance of z are found to be

$$E[z] = \int_0^\infty z \frac{1}{2^{n/2}\Gamma(n/2)} z^{n/2-1} e^{-z/2} \, dz = n, \tag{2.36}$$

$$V[z] = \int_0^\infty (z-n)^2 \frac{1}{2^{n/2}\Gamma(n/2)} z^{n/2-1} e^{-z/2} \, dz = 2n. \tag{2.37}$$

The χ^2 distribution is shown in Fig. 2.7 for several values of the parameter n.

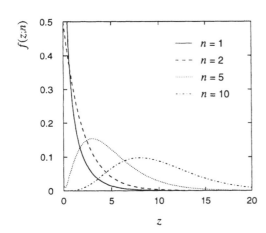

Fig. 2.7 The χ^2 probability density for various values of the parameter n.

The χ^2 distribution derives its importance from its relation to the sum of squares of Gaussian distributed variables. Given N independent Gaussian random variables x_i with known mean μ_i and variance σ_i^2, the variable

$$z = \sum_{i=1}^{N} \frac{(x_i - \mu_i)^2}{\sigma_i^2} \tag{2.38}$$

is distributed according to the χ^2 distribution for N degrees of freedom (see Section 10.2). More generally, if the x_i are not independent but are described by an N-dimensional Gaussian p.d.f. (equation (2.28)), the variable

$$z = (\mathbf{x} - \boldsymbol{\mu})^T V^{-1} (\mathbf{x} - \boldsymbol{\mu}) \tag{2.39}$$

is a χ^2 random variable for N degrees of freedom. Variables following the χ^2 distribution play an important role in tests of goodness-of-fit (Sections 4.5, 4.7), especially in conjunction with the method of least squares (Section 7.5).

2.8   Cauchy (Breit–Wigner) distribution

The **Cauchy** or **Breit–Wigner** p.d.f. of the continuous variable x $(-\infty < x < \infty)$ is defined by

$$f(x) = \frac{1}{\pi} \frac{1}{1 + x^2}. \tag{2.40}$$

This is a special case of the Breit–Wigner distribution encountered in particle physics,

$$f(x; \Gamma, x_0) = \frac{1}{\pi} \frac{\Gamma/2}{\Gamma^2/4 + (x - x_0)^2}, \tag{2.41}$$

where the parameters x_0 and Γ correspond to the mass and width of a resonance particle. This is shown in Fig. 2.8 for several values of the parameters.

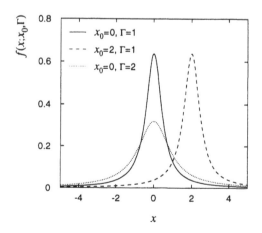

Fig. 2.8 The Cauchy (Breit–Wigner) probability density for various values of the parameters x_0 and Γ.

The expectation value of the Cauchy distribution is not well defined, since although the p.d.f. is symmetric about zero (or x_0 for (2.41)) the integrals

$\int_{-\infty}^{0} x f(x) dx$ and $\int_{0}^{\infty} x f(x) dx$ are individually divergent. The variance and higher moments are also divergent. The parameters x_0 and Γ can nevertheless be used to give information about the position and width of the p.d.f., as can be seen from the figure; x_0 is the peak position (i.e. the mode) and Γ is the full-width of the peak at half of the maximum height.[1]

2.9   Landau distribution

In nuclear and particle physics one often encounters the probability density $f(\Delta; \beta)$ for the energy loss Δ of a charged particle when traversing a layer of matter of a given thickness. This was first derived by Landau [Lan44], and is given by

$$f(\Delta; \beta) = \frac{1}{\xi}\, \phi(\lambda), \quad 0 \le \Delta < \infty, \tag{2.42}$$

where ξ is a parameter related to the properties of the material and the velocity of the particle, $\beta = v/c$, (measured in units of the velocity of light c) and $\phi(\lambda)$ is the p.d.f. of the dimensionless random variable λ. The variable λ is related to the properties of the material, the velocity β, and the energy loss Δ. These quantities are given by

$$\xi = \frac{2\pi N_A e^4 z^2 \rho \sum Z}{m_e c^2 \sum A} \frac{d}{\beta^2}, \tag{2.43}$$

$$\lambda = \frac{1}{\xi}\left[\Delta - \xi\left(\log\frac{\xi}{\epsilon'} + 1 - \gamma_E\right)\right], \tag{2.44}$$

$$\epsilon' = \frac{I^2 \exp(\beta^2)}{2 m_e c^2 \beta^2 \gamma^2}, \tag{2.45}$$

where N_A is Avagadro's number, m_e and e are the mass and charge of the electron, z is the charge of the incident particle in units of the electron's charge, $\sum Z$ and $\sum A$ are the sums of the atomic numbers and atomic weights of the molecular substance, ρ is its density, d is the thickness of the layer, $I = I_0 Z$ with $I_0 \approx 13.5$ eV is an ionization energy characteristic of the material, $\gamma = 1/\sqrt{1 - \beta^2}$, and $\gamma_E = 0.5772\ldots$ is Euler's constant. The function $\phi(\lambda)$ is given by

$$\phi(\lambda) = \frac{1}{2\pi i} \int_{\epsilon - i\infty}^{\epsilon + i\infty} \exp(u \log u + \lambda u)\, du, \tag{2.46}$$

where ϵ is infinitesimal and positive, or equivalently after a variable transformation by

---

[1]The definition used here is standard in high energy physics where Γ is interpreted as the decay rate of a particle. In some references, e.g. [Ead71, Fro79], the parameter Γ is defined as the half-width at half maximum, i.e. the p.d.f. is given by equation (2.41) with the replacement $\Gamma \to 2\Gamma$.

$$\phi(\lambda) = \frac{1}{\pi} \int_0^\infty \exp(-u \log u - \lambda u) \sin \pi u \, du. \tag{2.47}$$

The integral must be evaluated numerically, e.g. with the routine LANDAU [CER97, Koe84]. The energy loss distribution is shown in Fig. 2.9(a) for several values of the velocity $\beta = v/c$. Because of the long tail extending to high values of Δ, the mean and higher moments of the Landau distribution do not exist, i.e. the integral $\int_0^\infty \Delta^n f(\Delta) d\Delta$ diverges for $n \geq 1$. As can be seen from the figure, however, the most probable value (mode) Δ_{mp} is sensitive to the particle's velocity. This has been computed numerically in [Mac69] to be

$$\Delta_{mp} = \xi \left[\log(\xi/\epsilon') + 0.198 \right], \tag{2.48}$$

and is shown in Fig. 2.9(b).[2]

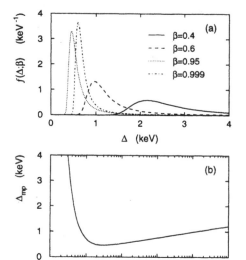

Fig. 2.9 (a) The Landau probability density for the energy loss Δ of a charged particle traversing a 4 mm thick layer of argon gas at standard temperature and pressure for various values of the velocity β. (b) The peak position (mode) of the distributions in (a) as a function of $\beta\gamma$ as given by equation (2.48).

Although the mean and higher moments do not exist for the Breit–Wigner and Landau distributions, the probability densities actually describing physical processes must have finite moments. If, for example, one were to measure the energy loss Δ of a particle in a particular system many times, the average would eventually converge to some value, since Δ cannot exceed the energy of the incoming particle. Similarly, the mass of a resonance particle cannot be less than the sum of the rest masses of its decay products, and it cannot be more than the center-of-mass energy of the reaction in which it was created. The problem arises because the Cauchy and Landau distributions are only approximate models of

---

[2]Equation (2.48) (the 'Bethe–Bloch formula') forms the basis for identification of charged particles by measurement of ionization energy loss, cf. [All80].

the physical system. The models break down in the tails of the distributions, which is the part of the p.d.f. that causes the mean and higher moments to diverge.

3

The Monte Carlo method

The Monte Carlo method is a numerical technique for calculating probabilities and related quantities by using sequences of random numbers. For the case of a single random variable, the procedure can be broken into the following stages. First, a series of random values $r_1, r_2 \ldots$ is generated according to a uniform distribution in the interval $0 < r < 1$. That is, the p.d.f. $g(r)$ is given by

$$g(r) = \begin{cases} 1 & 0 < r < 1, \\ 0 & \text{otherwise}. \end{cases} \tag{3.1}$$

Next, the sequence $r_1, r_2, \ldots$ is used to determine another sequence $x_1, x_2 \ldots$ such that the x values are distributed according to a p.d.f. $f(x)$ in which one is interested. The values of x can then be treated as simulated measurements, and from them the probabilities for x to take on values in a certain region can be estimated. In this way one effectively computes an integral of $f(x)$. This may seem like a trivial exercise, since the function $f(x)$ was available to begin with, and could simply have been integrated over the region of interest. The true usefulness of the technique, however, becomes apparent in multidimensional problems, where integration of a joint p.d.f. $f(x, y, z, \ldots)$ over a complicated region may not be feasible by other methods.

3.1 Uniformly distributed random numbers

In order to generate a sequence of uniformly distributed random numbers, one could in principle make use of a random physical process such as the repeated tossing of a coin. In practice, however, this task is almost always accomplished by a computer algorithm called a **random number generator**. Many such algorithms have been implemented as user-callable subprograms (e.g. the routines RANMAR [Mar91] or RANLUX [Lüs94, Jam94], both in [CER97]). A detailed discussion of random number generators is beyond the scope of this book and the interested reader is referred to the more complete treatments in [Bra92, Jam90]. Here a simple but effective algorithm will be presented in order to illustrate the general idea.

A commonly used type of random number generator is based on the **multiplicative linear congruential** algorithm. Starting from an initial integer value n_0 (called the **seed**), one generates a sequence of integers $n_1, n_2, \ldots$ according to the rule

$$n_{i+1} = (an_i) \bmod m. \tag{3.2}$$

Here the **multiplier** a and **modulus** m are integer constants and the mod (modulo) operator means that one takes the remainder of an_i divided by m. The values n_i follow a periodic sequence in the range $[1, m-1]$. In order to obtain values uniformly distributed in $(0, 1)$, one uses the transformation

$$r_i = n_i/m. \tag{3.3}$$

Note that this excludes 0 and 1; in some other algorithms these values can be included. The initial value n_0 and the two constants a and m determine the entire sequence, which, of course, is not truly random, but rather strictly determined. The resulting values are therefore more correctly called **pseudorandom**. For essentially all applications these can be treated as equivalent to true random numbers, with the exception of being reproducible, e.g. if one repeats the procedure with the same seed.

The values of m and a are chosen such that the generated numbers perform well with respect to various tests of randomness. Most important among these is a long period before the sequence repeats, since after this occurs the numbers can clearly no longer be regarded as random. In addition, one tries to attain the smallest possible correlations between pairs of generated numbers. For a 32-bit integer representation, for example, $m = 2147483399$ and $a = 40692$ have been shown to give good results, and with these one attains the maximum period of $m - 1 \approx 2 \times 10^9$ [Lec88]. More sophisticated algorithms allow for much longer periods, e.g. approximately 10^{43} for the RANMAR generator [Mar91, CER97].

3.2 The transformation method

Given a sequence of random numbers $r_1, r_2, \ldots$ uniformly distributed in $[0, 1]$, the next step is to determine a sequence $x_1, x_2, \ldots$ distributed according to the p.d.f. $f(x)$ in which one is interested. In the transformation method this is accomplished by finding a suitable function $x(r)$ which directly yields the desired sequence when evaluated with the uniformly generated r values. The problem is clearly related to the transformation of variables discussed in Section 1.4. There, an original p.d.f. $f(x)$ for a random variable x and a function $a(x)$ were specified, and the p.d.f. $g(a)$ for the function a was then found. Here the task is to find a function $x(r)$ that is distributed according to a specified $f(x)$, given that r follows a uniform distribution between 0 and 1.

The probability to obtain a value of r in the interval $[r, r+dr]$ is $g(r)dr$, and this should be equal to the probability to obtain a value of x in the corresponding interval $[x(r), x(r)+dx(r)]$, which is $f(x)dx$. In order to determine $x(r)$ such that this is true, one can require that the probability that r is less than some value r' be equal to the probability that x is less than $x(r')$. (We will see in the following example that this prescription is not unique.) That is, one must find a function $x(r)$ such that $F(x(r)) = G(r)$, where F and G are the cumulative distributions corresponding to the p.d.f.s f and g. Since the cumulative distribution for the uniform p.d.f. is $G(r) = r$ with $0 \leq r \leq 1$, one has

$$F(x(r)) = \int_{-\infty}^{x(r)} f(x')dx' \quad = \quad \int_{-\infty}^{r} g(r')dr'$$

$$= \quad r. \tag{3.4}$$

Equation (3.4) says in effect that the cumulative distribution $F(x)$, treated as a random variable, is uniformly distributed between 0 and 1 (cf. equation (2.18)).

Depending on the $f(x)$ in question, it may or may not be possible to solve for $x(r)$ using equation (3.4). Consider the exponential distribution discussed in Section 2.4. Equation (3.4) becomes

$$\int_{0}^{x(r)} \frac{1}{\xi} e^{-x'/\xi} dx' = r. \tag{3.5}$$

Integrating and solving for x gives

$$x(r) = -\xi \log(1 - r). \tag{3.6}$$

If the variable r is uniformly distributed between 0 and 1 then $r' = 1 - r$ clearly is too, so that the function

$$x(r) = -\xi \log r \tag{3.7}$$

also has the desired property. That is, if r follows a uniform distribution between 0 and 1, then $x(r) = -\xi \log r$ will follow an exponential distribution with mean ξ.

3.3   The acceptance–rejection method

It turns out to be too difficult in many practical applications to solve equation (3.4) for $x(r)$ analytically. A useful alternative is von Neumann's acceptance–rejection technique [Neu51]. Consider a p.d.f. $f(x)$ which can be completely surrounded by a box between x_{min} and x_{max} and having height f_{max}, as shown in Fig. 3.1. One can generate a series of numbers distributed according to $f(x)$ with the following algorithm:

(1) Generate a random number x, uniformly distributed between x_{min} and x_{max}, i.e. $x = x_{min} + r_1(x_{max} - x_{min})$ where r_1 is uniformly distributed between 0 and 1.

(2) Generate a second independent random number u uniformly distributed between 0 and f_{max}, i.e. $u = r_2 f_{max}$.

(3) If $u < f(x)$, then accept x. If not, reject x and repeat.

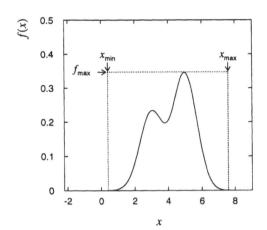

Fig. 3.1 A probability density $f(x)$ enclosed by a box to generate random numbers using the acceptance-rejection technique.

The accepted x values will be distributed according to $f(x)$, since for each value of x obtained from step (1) above, the probability to be accepted is proportional to $f(x)$.

As an example consider the p.d.f.[1]

$$f(x) = \frac{3}{8}(1 + x^2), \quad -1 \le x \le 1. \tag{3.8}$$

At $x = \pm 1$ the p.d.f. has a maximum value of $f_{max} = 3/4$. Figure 3.2(a) shows a scatter plot of the random numbers u and x generated according to the algorithm given above. The x values of the points that lie below the curve are accepted. Figure 3.2(b) shows a normalized histogram constructed from the accepted points.

The efficiency of the algorithm (i.e. the fraction of x values accepted) is the ratio of the areas of the p.d.f. (unity) to that of the enclosing box $f_{max} \cdot (x_{max} - x_{min})$. For a highly peaked density function the efficiency may be quite low, and the algorithm may be too slow to be practical. In cases such as these, one can improve the efficiency by enclosing the p.d.f. $f(x)$ in any other curve $g(x)$ for which random numbers can be generated according to $g(x)/\int g(x')dx'$, using, for example, the transformation method.

The more general algorithm is then:

(1) Generate a random number x according to the p.d.f. $g(x)/\int g(x')dx'$.
(2) Generate a second random number u uniformly distributed between 0 and $g(x)$.
(3) If $u < f(x)$, then accept x. If not, reject x and repeat.

---

[1]Equation (3.8) gives the distribution of the scattering angle θ in the reaction $e^+e^- \to \mu^+\mu^-$ with $x = \cos\theta$ (see e.g. [Per87]).

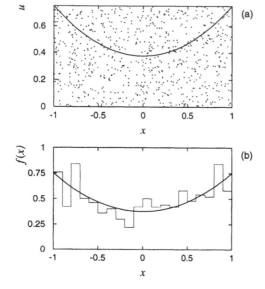

Fig. 3.2 (a) Scatter plot of pairs of numbers (u, x), where x is uniformly distributed in $-1 \leq x \leq 1$, and u is uniform in $0 \leq u \leq f_{\max}$. The x values of the points below the curve are accepted. (b) Normalized histogram of the accepted x values with the corresponding p.d.f.

Here the probability to generate a value x in step (1) is proportional to $g(x)$, and the probability to be retained after step (3) is equal to $f(x)/g(x)$, so that the total probability to obtain x is proportional to $f(x)$ as required.

3.4   Applications of the Monte Carlo method

The Monte Carlo method can be applied whenever the solution to a problem can be related to a parameter of a probability distribution. This could be either an explicit parameter in a p.d.f., or the integral of the distribution over some region. A sequence of Monte Carlo generated values is used to evaluate an estimator for the parameter (or integral), just as is done with real data. Techniques for constructing estimators are discussed in Chapters 5–8.

An important feature of properly constructed estimators is that their statistical accuracy improves as the number of values n in the data sample (from Monte Carlo or otherwise) increases. One can show that under fairly general conditions, the standard deviation of an estimator is inversely proportional to $\sqrt{n}$ (see Section 6.6). The Monte Carlo method thus represents a numerical integration technique for which the accuracy increases as $1/\sqrt{n}$.

This scaling behavior with the number of generated values can be compared to the number of points necessary to compute an integral using the trapezoidal rule. Here the accuracy improves as $1/n^2$, i.e. much faster than by Monte Carlo. For an integral of dimension d, however, this is changed to $1/n^{2/d}$, whereas for Monte Carlo integration one has $1/\sqrt{n}$ for any dimension. So for $d > 4$, the dependence of the accuracy on n is better for the Monte Carlo method. For other integration methods, such as Gaussian quadrature, a somewhat better rate of convergence can be achieved than for the trapezoidal rule. For a large enough number of dimensions, however, the Monte Carlo method will always be superior. A more detailed discussion of these considerations can be found in [Jam80].

The Monte Carlo technique provides a method for determining the p.d.f.s of functions of random variables. Suppose, for example, one has n independent random variables $x_1, \ldots, x_n$ distributed according to known p.d.f.s $f_1(x_1)$, $\ldots$, $f_n(x_n)$, and one would like to compute the p.d.f. $g(a)$ of some (possibly complicated) function $a(x_1, \ldots, x_n)$. The techniques described in Section 1.4 are often only usable for relatively simple functions of a small number of variables. With the Monte Carlo method, a value for each x_i is generated according to the corresponding $f_i(x_i)$. The value of $a(x_1, \ldots, x_n)$ is then computed and recorded (e.g. in a histogram). The procedure is repeated until one has enough values of a to estimate the properties of its p.d.f. $g(a)$ (e.g. mean, variance) with the desired statistical precision. Examples of this technique will be used in the following chapters.

The Monte Carlo method is often used to simulate experimental data. In particle physics, for example, this is typically done in two stages: event generation and detector simulation. Consider, for example, an experiment in which an incoming particle such as an electron scatters off a target and is then detected. Suppose there exists a theory that predicts the probability for an event to occur as a function of the scattering angle (i.e. the differential cross section). First one constructs a Monte Carlo program to generate values of the scattering angles, and thus the momentum vectors, of the final state particles. Such a program is called an **event generator**. In high energy physics, event generators are available to describe a wide variety of particle reactions.

The output of the event generator, i.e. the momentum vectors of the generated particles, is then used as input for a **detector simulation program**. Since the response of a detector to the passage of the scattered particles also involves random processes such as the production of ionization, multiple Coulomb scattering, etc., the detector simulation program is also implemented using the Monte Carlo method. Programming packages such as GEANT [CER97] can be used to describe complicated detector configurations, and experimental collaborations typically spend considerable effort in achieving as complete a modeling of the detector as possible. This is especially important in order to optimize the detector's design for investigating certain physical processes before investing time and money in constructing the apparatus.

When the Monte Carlo method is used to simulate experimental data, one can most easily think of the procedure as a computer implementation of an intrinsically random process. Probabilities are naturally interpreted as relative frequencies of outcomes of a repeatable experiment, and the experiment is simply repeated many times on the computer. The Monte Carlo method can also be regarded, however, as providing a numerical solution to any problem that can be related to probabilities. The results are clearly independent of the probability interpretation. This is the case, for example, when the Monte Carlo method is used simply to carry out a transformation of variables or to compute integrals of functions which may not normally be interpreted as probability densities.

4

Statistical tests

In this chapter some basic concepts of statistical test theory are presented. As this is a broad topic, after a general introduction we will limit the discussion to several aspects that are most relevant to particle physics. Here one could be interested, for example, in the particles resulting from an interaction (an event), or one might consider an individual particle within an event. An immediate application of statistical tests in this context is the selection of candidate particles or events which are then used for further analysis. Here one is concerned with distinguishing events of interest (signal) from other types (background). These questions are addressed in Sections 4.2–4.4. Another important aspect of statistical tests concerns goodness-of-fit; this is discussed in Sections 4.5–4.7.

4.1 Hypotheses, test statistics, significance level, power

The goal of a statistical test is to make a statement about how well the observed data stand in agreement with given predicted probabilities, i.e. a **hypothesis**. The hypothesis under consideration is traditionally called the **null hypothesis**, H_0, which could specify, for example, a probability density $f(x)$ of a random variable x. If the hypothesis determines $f(x)$ uniquely it is said to be **simple**; if the form of the p.d.f. is defined but not the values of at least one free parameter θ, then $f(x; \theta)$ is called a **composite hypothesis**. In such cases the unknown parameter or parameters are estimated from the data using, say, techniques discussed in Chapters 5–8. For now we will concentrate on simple hypotheses.

A statement about the validity of H_0 often involves a comparison with some **alternative hypotheses**, $H_1, H_2, \ldots$. Suppose one has data consisting of n measured values $\mathbf{x} = (x_1, \ldots, x_n)$, and a set of hypotheses, $H_0, H_1, \ldots$, each of which specifies a given joint p.d.f., $f(\mathbf{x}|H_0), f(\mathbf{x}|H_1), \ldots$.[1] The values could, for example, represent n repeated observations of the same random variable, or a single observation of an n-dimensional variable. In order to investigate the measure of agreement between the observed data and a given hypothesis, one constructs a function of the measured variables called a **test statistic** $t(\mathbf{x})$. Each of the hypotheses will imply a given p.d.f. for the statistic t, i.e. $g(t|H_0), g(t|H_1)$, etc.

---

[1] For the p.d.f. of $\mathbf{x}$ given the hypothesis H the notation of conditional probability $f(\mathbf{x}|H)$ is used (Section 1.3), even though in the context of classical statistics a hypothesis H is only regarded as a random variable if it refers to the outcome of a repeatable experiment. In Bayesian statistics both $\mathbf{x}$ and H are random variables, so there the notation is in any event appropriate.

The test statistic t can be a multidimensional vector, $\mathbf{t} = (t_1, \ldots, t_m)$. In fact, the original vector of data values $\mathbf{x} = (x_1, \ldots, x_n)$ could be used. The point of constructing a statistic t of lower dimension (i.e. $m < n$) is to reduce the amount of data without losing the ability to discriminate between hypotheses. Let us suppose for the moment that we have chosen a scalar function $t(\mathbf{x})$, which has the p.d.f. $g(t|H_0)$ if H_0 is true, and $g(t|H_1)$ if H_1 is true, as shown in Fig. 4.1.

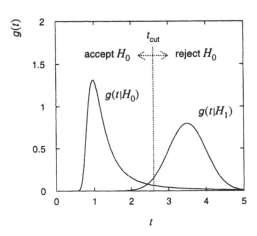

Fig. 4.1 Probability densities for the test statistic t under assumption of the hypotheses H_0 and H_1. H_0 is rejected if t is observed in the critical region, here shown as $t > t_{\text{cut}}$.

Often one formulates the statement about the compatibility between the data and the various hypotheses in terms of a decision to accept or reject a given null hypothesis H_0. This is done by defining a **critical region** for t. Equivalently, one can use its complement, called the **acceptance region**. If the value of t actually observed is in the critical region, one rejects the hypothesis H_0; otherwise, H_0 is accepted. The critical region is chosen such that the probability for t to be observed there, under assumption of the hypothesis H_0, is some value α, called the **significance level** of the test. For example, the critical region could consist of values of t greater than a certain value t_{cut}, called the **cut** or **decision boundary**, as shown in Fig. 4.1. The significance level is then

$$\alpha = \int_{t_{\text{cut}}}^{\infty} g(t|H_0)\,dt. \tag{4.1}$$

One would then accept (or, strictly speaking, not reject) the hypothesis H_0 if the value of t observed is less than t_{cut}. There is thus a probability of α to reject H_0 if H_0 is true. This is called an **error of the first kind**. An **error of the second kind** takes place if the hypothesis H_0 is accepted (i.e. t is observed less than t_{cut}) but the true hypothesis was not H_0 but rather some alternative hypothesis H_1. The probability for this is

$$\beta = \int_{-\infty}^{t_{\text{cut}}} g(t|H_1)\,dt. \tag{4.2}$$

where $1 - \beta$ is called the **power** of the test to discriminate against the alternative hypothesis H_1.

4.2   An example with particle selection

As an example, the test statistic t could represent the measured ionization created by a charged particle of a known momentum traversing a detector. The amount of ionization is subject to fluctuations from particle to particle, and depends (for a fixed momentum) on the particle's mass. Thus the p.d.f. $g(t|H_0)$ in Fig. 4.1 could correspond to the hypothesis that the particle is an electron, and the $g(t|H_1)$ could be what one would obtain if the particle was a pion, i.e. $H_0 = $ e, $H_1 = \pi$.

Suppose the particles in question are all known to be either electrons or pions, and that one would like to select a sample of electrons by requiring $t \leq t_{\text{cut}}$. (The electrons are regarded as 'signal', and pions are considered as 'background'.) The probabilities to accept a particle of a given type, i.e. the **selection efficiencies** ε_e and ε_π, are thus

$$\varepsilon_e = \int_{-\infty}^{t_{\text{cut}}} g(t|e)dt = 1 - \alpha, \qquad (4.3)$$

$$\varepsilon_\pi = \int_{-\infty}^{t_{\text{cut}}} g(t|\pi)dt = \beta. \qquad (4.4)$$

Individually these can be made arbitrarily close to zero or unity simply by an appropriate choice of the critical region, i.e. by making a looser or tighter cut on the ionization. The price one pays for a high efficiency for the signal is clearly an increased amount of contamination, i.e. the purity of the electron sample decreases because some pions are accepted as well.

If the relative fractions of pions and electrons are not known, the problem becomes one of parameter estimation (Chapters 5–8). That is, the test statistic t will be distributed according to

$$f(t; a_e) = a_e g(t|e) + (1 - a_e)g(t|\pi), \qquad (4.5)$$

where a_e and $a_\pi = 1 - a_e$ are the fractions of electrons and pions, respectively. An estimate of a_e then gives the total number of electrons N_e in the original sample of N_{tot} particles, $N_e = a_e N_{\text{tot}}$.

Alternatively one may want to select a set of electron candidates by requiring $t < t_{\text{cut}}$, leading to N_{acc} accepted out of the N_{tot} particles. One is then often interested in determining the total number of electrons present before the cut on t was made. The number of accepted particles is given by[2]

---

[2] Strictly speaking, equations (4.6) and (4.7) give expectation values of numbers of particles, not the numbers which would necessarily be found in an experiment with a finite data sample. The distinction will be dealt with in more detail in Chapter 5.

$$N_{\text{acc}} = \varepsilon_e N_e + \varepsilon_\pi N_\pi$$
$$= \varepsilon_e N_e + \varepsilon_\pi (N_{\text{tot}} - N_e), \tag{4.6}$$

which gives

$$N_e = \frac{N_{\text{acc}} - \varepsilon_\pi N_{\text{tot}}}{\varepsilon_e - \varepsilon_\pi}. \tag{4.7}$$

From (4.7) one sees that the number of accepted particles N_{acc} can only be used to determine the number of electrons N_e if the efficiencies ε_e and ε_π are different. If there are uncertainties in ε_π and ε_e, then these will translate into an uncertainty in N_e according to the error propagation techniques of Section 1.6. One would try to select the critical region (i.e. the cut value for the ionization) in such a way that the total error in N_e is a minimum.

The probabilities that a particle with an observed value of t is an electron or a pion, $h(e|t)$ and $h(\pi|t)$, are obtained from the p.d.f.s $g(t|e)$ and $g(t|\pi)$ using Bayes' theorem (1.8),

$$h(e|t) = \frac{a_e\, g(t|e)}{a_e\, g(t|e) + a_\pi\, g(t|\pi)}, \tag{4.8}$$

$$h(\pi|t) = \frac{a_\pi\, g(t|\pi)}{a_e\, g(t|e) + a_\pi\, g(t|\pi)}, \tag{4.9}$$

where a_e and $a_\pi = 1 - a_e$ are the prior probabilities for the hypotheses e and π. Thus in order to give the probability that a given selected particle is an electron, one needs the prior probabilities for all of the possible hypotheses as well as the p.d.f.s that they imply for the statistic t.

Although this is essentially the Bayesian approach to the problem, equations (4.8) and (4.9) also make sense in the framework of classical statistics. If one is dealing with a large sample of particles, then the hypotheses $H = e$ and $H = \pi$ refer to a characteristic that changes randomly from particle to particle. Using the relative frequency interpretation in this case, $h(e|t)$ gives the fraction of times a particle with a given t will be an electron. In Bayesian statistics using subjective probability, one would say that $h(e|t)$ gives the degree of belief that a given particle with a measured value of t is an electron.

Instead of the probability that an individual particle is an electron, one may be interested in the **purity** p_e of a sample of electron candidates selected by requiring $t < t_{\text{cut}}$. This is given by

$$p_e = \frac{\text{number of electrons with } t < t_{\text{cut}}}{\text{number of all particles with } t < t_{\text{cut}}}$$

$$= \frac{\int_{-\infty}^{t_{\text{cut}}} a_e g(t|e) dt}{\int_{-\infty}^{t_{\text{cut}}} (a_e g(t|e) + (1 - a_e) g(t|\pi)) dt}$$

$$= \frac{a_e \varepsilon_e N_{\text{tot}}}{N_{\text{acc}}}. \tag{4.10}$$

One can check using equation (4.8) that this is simply the mean electron probability $h(e|t)$, averaged over the interval $(-\infty, t_{\text{cut}}]$. That is,

$$p_e = \frac{\int_{-\infty}^{t_{\text{cut}}} h(e|t) f(t) dt}{\int_{-\infty}^{t_{\text{cut}}} f(t) dt}. \tag{4.11}$$

4.3 Choice of the critical region using the Neyman–Pearson lemma

Up to now the exact choice of the critical region, i.e. the value of t_{cut}, was left open. This will be chosen depending on the efficiency and purity of the selected particles (or events) desired in the further analysis. One way of defining an optimal placement of the cuts is to require that they give a maximum purity for a given efficiency. (The desired value of the efficiency is still left open.) With the one-dimensional test statistic of the previous example, this was achieved automatically. There only a single cut value t_{cut} needed to be determined, and this determined both the efficiency and purity.

Suppose, however, that a multidimensional test statistic $\mathbf{t} = (t_1, \ldots, t_m)$ has been chosen. The definition of the critical (or acceptance) region is then not as obvious. Assume we wish to test a simple hypothesis H_0, say, in order to select events of a given type. We will allow for a simple alternative hypothesis H_1. That is, in addition to the events of interest (signal) there are also background events, so that the signal purity in the selected sample will in general be less than 100%.

The **Neyman–Pearson lemma** states that the acceptance region giving the highest power (and hence the highest signal purity) for a given significance level α (or selection efficiency $\varepsilon = 1 - \alpha$) is the region of $\mathbf{t}$-space such that

$$\frac{g(\mathbf{t}|H_0)}{g(\mathbf{t}|H_1)} > c. \tag{4.12}$$

Here c is a constant which is determined by the desired efficiency. A proof can be found in [Bra92]. Note that a test based on the Neyman–Pearson acceptance region for the vector statistic $\mathbf{t}$ is in fact equivalent to a test using a one-dimensional statistic given by the ratio on the left-hand side of (4.12),

$$r = \frac{g(\mathbf{t}|H_0)}{g(\mathbf{t}|H_1)}. \tag{4.13}$$

This is called the **likelihood ratio** for simple hypotheses H_0 and H_1. The corresponding acceptance region is given by $r > c$.

4.4   Constructing a test statistic

Suppose that we start with a vector of data $\mathbf{x} = (x_1, \ldots, x_n)$, and we would like to construct out of this a one-dimensional test statistic $t(\mathbf{x})$ so as to distinguish between two simple hypotheses H_0 and H_1. As we have seen in the previous section, the best test statistic in the sense of maximum power (and hence maximum signal purity) for a given significance level (or selection efficiency) is given by the likelihood ratio,

$$t(\mathbf{x}) = \frac{f(\mathbf{x}|H_0)}{f(\mathbf{x}|H_1)}. \tag{4.14}$$

In order to construct this, however, we need to know $f(\mathbf{x}|H_0)$ and $f(\mathbf{x}|H_1)$. Often these must be determined by a Monte Carlo simulation of the two types of events, where the probability densities are represented as multidimensional histograms. If one has M bins for each of the n components of $\mathbf{x}$, then the total number of bins is M^n. This means in effect that M^n parameters must be determined from the Monte Carlo data. For n too large, the method becomes impractical because of the prohibitively large number of events needed.

Even if we cannot determine $f(\mathbf{x}|H_0)$ and $f(\mathbf{x}|H_1)$ as n-dimensional histograms, we can nevertheless make a simpler *Ansatz* for the functional form of a test statistic $t(\mathbf{x})$, and then choose the best function (according to some criteria) having this form. We will consider both linear and nonlinear functions of the x_i. In addition, we will need to address some practical considerations concerning the choice of the input variables.

4.4.1   Linear test statistics, the Fisher discriminant function

The simplest form for the statistic $t(\mathbf{x})$ is a linear function,

$$t(\mathbf{x}) = \sum_{i=1}^{n} a_i x_i = \mathbf{a}^T \mathbf{x}, \tag{4.15}$$

where $\mathbf{a}^T = (a_1, \ldots, a_n)$ is the transpose (i.e. row) vector of coefficients. The goal is to determine the a_i so as to maximize the separation between the p.d.f.s $g(t|H_0)$ and $g(t|H_1)$. Different definitions of the separation will lead to different rules for determining the coefficients. One approach, first developed by Fisher [Fis36], is based on the following considerations. The data $\mathbf{x}$ have the mean values and covariance matrix

$$
\begin{aligned}
(\mu_k)_i &= \int x_i f(\mathbf{x}|H_k) \, dx_1 \ldots dx_n, \\
(V_k)_{ij} &= \int (x - \mu_k)_i \, (x - \mu_k)_j \, f(\mathbf{x}|H_k) \, dx_1 \ldots dx_n,
\end{aligned}
\tag{4.16}
$$

where the indices $i, j = 1, \ldots, n$ refer to the components of the vector $\mathbf{x}$, and $k = 0, 1$ refers to hypotheses H_0 or H_1. In a corresponding way, each hypothesis is characterized by a certain expectation value and variance for t,

$$
\begin{aligned}
\tau_k &= \int t g(t|H_k)\, dt = \mathbf{a}^T \boldsymbol{\mu}_k, \\
\Sigma_k^2 &= \int (t - \tau_k)^2 g(t|H_k)\, dt = \mathbf{a}^T V_k\, \mathbf{a}.
\end{aligned}
\tag{4.17}
$$

To increase the separation we should clearly try to maximize $|\tau_0 - \tau_1|$. In addition, we want $g(t|H_0)$ and $g(t|H_1)$ to be as tightly concentrated as possible about τ_0 and τ_1; this is determined by the variances Σ_0^2 and Σ_1^2. A measure of separation that takes both of these considerations into account is

$$
J(\mathbf{a}) = \frac{(\tau_0 - \tau_1)^2}{\Sigma_0^2 + \Sigma_1^2}.
\tag{4.18}
$$

Expressing the numerator in terms of the a_i, one finds

$$
\begin{aligned}
(\tau_0 - \tau_1)^2 &= \sum_{i,j=1}^{n} a_i a_j (\mu_0 - \mu_1)_i (\mu_0 - \mu_1)_j \\
&= \sum_{i,j=1}^{n} a_i a_j B_{ij} = \mathbf{a}^T B\, \mathbf{a},
\end{aligned}
\tag{4.19}
$$

where the matrix B, defined as

$$
B_{ij} = (\mu_0 - \mu_1)_i\, (\mu_0 - \mu_1)_j \quad i, j = 1, \ldots, n,
\tag{4.20}
$$

represents the separation 'between' the two classes corresponding to H_0 and H_1. Similarly, the denominator of (4.18) becomes

$$
\Sigma_0^2 + \Sigma_1^2 = \sum_{i,j=1}^{n} a_i a_j (V_0 + V_1)_{ij} = \mathbf{a}^T W\, \mathbf{a},
\tag{4.21}
$$

where $W_{ij} = (V_0 + V_1)_{ij}$ represents the sum of the covariance matrices 'within' the classes. The measure of separation (4.18) thus becomes

$$
J(\mathbf{a}) = \frac{\mathbf{a}^T B\, \mathbf{a}}{\mathbf{a}^T W\, \mathbf{a}}.
\tag{4.22}
$$

Setting the derivatives of $J(\mathbf{a})$ with respect to the a_i equal to zero to find the maximum separation gives

$$\mathbf{a} \propto W^{-1} \left(\boldsymbol{\mu}_0 - \boldsymbol{\mu}_1 \right). \tag{4.23}$$

The coefficients are only determined up to an arbitrary scale factor. The test statistic based on (4.15) and (4.23) is called **Fisher's linear discriminant function**. In order to determine the coefficients a_i, one needs the matrix W and the expectation values $\boldsymbol{\mu}_{(0,1)}$. These are usually estimated from a set of training data, e.g. from a Monte Carlo simulation. The important point is that one does not need to determine the full joint p.d.f.s $f(\mathbf{x}|H_0)$ and $f(\mathbf{x}|H_1)$ as n-dimensional histograms, but rather only the means and covariances (4.16) must be found. That is, the training data are used to determine only $n(n+1)/2$ parameters. Functions used to estimate means and covariances from a data sample are given in Section 5.2.

It is possible to change the scale of the variable t simply by multiplying $\mathbf{a}$ by a constant. By generalizing the definition of t to read

$$t(\mathbf{x}) = a_0 + \sum_{i=1}^{n} a_i x_i, \tag{4.24}$$

one can use the offset a_0 and the arbitrary scale to fix the expectation values τ_0 and τ_1 to any desired values. Maximizing the class separation (4.18) with fixed τ_0 and τ_1 is then equivalent to minimizing the sum of the variances within the classes,

$$\Sigma_0^2 + \Sigma_1^2 = E_0[(t - \tau_0)^2] + E_1[(t - \tau_1)^2], \tag{4.25}$$

where E_0 and E_1 denote the expectation values under the hypotheses H_0 and H_1. Thus the criterion used to determine the coefficients $\mathbf{a}$ is similar to a minimum principle that we will encounter in Chapter 7 concerning parameter estimation, namely the principle of least squares.

It is interesting to note a few properties of the Fisher discriminant for the case where the p.d.f.s $f(\mathbf{x}|H_0)$ and $f(\mathbf{x}|H_1)$ are both multidimensional Gaussians with common covariance matrices, $V_0 = V_1 = V$,

$$f(\mathbf{x}|H_k) = \frac{1}{(2\pi)^{n/2}|V|^{1/2}} \exp\left[-\tfrac{1}{2}(\mathbf{x} - \boldsymbol{\mu}_k)^T V^{-1}(\mathbf{x} - \boldsymbol{\mu}_k)\right], \ k = 0, 1. \tag{4.26}$$

In this case the Fisher discriminant, including an offset as in equation (4.24), can be taken to be

$$t(\mathbf{x}) = a_0 + (\boldsymbol{\mu}_0 - \boldsymbol{\mu}_1)^T V^{-1} \mathbf{x}. \tag{4.27}$$

The likelihood ratio (4.14) becomes

$$
\begin{aligned}
r &= \exp\left[-\tfrac{1}{2}(\mathbf{x} - \boldsymbol{\mu}_0)^T V^{-1}(\mathbf{x} - \boldsymbol{\mu}_0) + \tfrac{1}{2}(\mathbf{x} - \boldsymbol{\mu}_1)^T V^{-1}(\mathbf{x} - \boldsymbol{\mu}_1)\right] \\
&= \exp\left[(\boldsymbol{\mu}_0 - \boldsymbol{\mu}_1)^T V^{-1}\mathbf{x} - \tfrac{1}{2}\boldsymbol{\mu}_0^T V^{-1}\boldsymbol{\mu}_0 + \tfrac{1}{2}\boldsymbol{\mu}_1^T V^{-1}\boldsymbol{\mu}_1\right] \\
&\propto e^t.
\end{aligned}
\tag{4.28}
$$

That is, we have $t \propto \log r + \text{const.}$, and hence the test statistic (4.27) is given by a monotonic function of r. So for this case the Fisher discriminant function is just as good as the likelihood ratio. In addition, the multidimensional Gaussian with $V_0 = V_1 = V$ results in a simple expression for the posterior probabilities for the hypotheses. From Bayes' theorem we have for, say, the probability of H_0 given the data $\mathbf{x}$,

$$
P(H_0|\mathbf{x}) = \frac{f(\mathbf{x}|H_0)\pi_0}{f(\mathbf{x}|H_0)\pi_0 + f(\mathbf{x}|H_1)\pi_1} = \frac{1}{1 + \frac{\pi_1}{\pi_0 r}},
\tag{4.29}
$$

where π_0 and π_1 are the prior probabilities for H_0 and H_1. Combining this with the expression for r (4.28) gives

$$
P(H_0|\mathbf{x}) = \frac{1}{1 + e^{-t}} \equiv s(t),
\tag{4.30}
$$

and for the offset one obtains

$$
a_0 = -\tfrac{1}{2}\boldsymbol{\mu}_0^T V^{-1}\boldsymbol{\mu}_0 + \tfrac{1}{2}\boldsymbol{\mu}_1^T V^{-1}\boldsymbol{\mu}_1 + \log\frac{\pi_0}{\pi_1}.
\tag{4.31}
$$

The function $s(t)$ is a special case of the **sigmoid function**. The form used here takes on values in the interval $(0, 1)$, and is therefore called a **logistic sigmoid**.

4.4.2 Nonlinear test statistics, neural networks

If the joint p.d.f.s $f(\mathbf{x}|H_0)$ and $f(\mathbf{x}|H_1)$ are not Gaussian or if they do not have a common covariance matrix, then the Fisher discriminant no longer has the optimal properties seen above. One can then try a more general parametrization for the test statistic $t(\mathbf{x})$. Here we will consider a functional form which is a special case of an **artificial neural network**. The field of neural networks has developed a vast literature in recent years, especially for problems related to pattern recognition; here we will only sketch some of the main ideas. The interested reader is referred to more complete treatments such as [Bis95, Her91, Mül95]. Some applications of neural networks to problems in particle physics can be found in [Lön92, Pet92, Bab93].

Suppose we take $t(\mathbf{x})$ to be of the form

$$
t(\mathbf{x}) = s\left(a_0 + \sum_{i=1}^{n} a_i x_i\right).
\tag{4.32}
$$

(a)

(b)

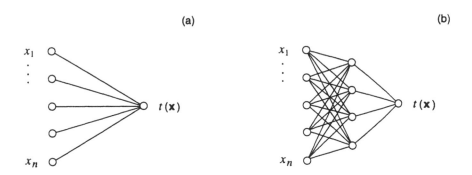

Fig. 4.2 Architectures of two feed-forward neural networks: (a) a single-layer perceptron and (b) a two-layer perceptron.

The function $s(\cdot)$ is in general called the **activation function**, which we will take to be a logistic sigmoid. Other activation functions such as a step function are also possible. The argument of the activation function is a linear function of the input variables, including the offset a_0 (often called a threshold). A test statistic of this form is called a **single-layer perceptron**.

The structure of the single-layer perceptron is illustrated in Fig. 4.2(a). The input values $\mathbf{x}$ are represented as a set of **nodes**, which together constitute the **input layer**. The final test statistic $t(\mathbf{x})$ is given by the **output node**. In general we could consider multiple output nodes, corresponding to a vector test statistic. As in the preceding section we will restrict ourselves here to a single output node.

Since the sigmoid function is monotonic, the single-layer perceptron is equivalent to a linear test statistic. This can now be generalized, however, to the **two-layer perceptron** in the manner illustrated in Fig. 4.2(b). In addition to the input layer, one has a **hidden layer** with m nodes, $(h_1, \ldots, h_m)$. The h_i take over the role of the input variables of the single-layer perceptron, so that t is now given by

$$t(\mathbf{x}) = s \left(a_0 + \sum_{i=1}^{m} a_i h_i(\mathbf{x}) \right).$$ (4.33)

The h_i themselves are given as functions of the nodes in the previous layer (here, the input layer),

$$h_i(\mathbf{x}) = s \left(w_{i0} + \sum_{i=1}^{n} w_{ij} x_j \right).$$ (4.34)

This can easily be generalized to an arbitrary number of hidden layers (the multilayer perceptron). One usually restricts the connections so that the value of a given node only depends on the nodes in the previous layer, as indicated in Fig. 4.2; this is called a **feed-forward network**. The number of free parameters for n input variables followed by layers containing m_1, m_2, m_3, ... nodes is then given by $(n+1)m_1 + (m_1+1)m_2 + (m_2+1)m_3 + \cdots$. (Note that this includes the a_0, w_{i0}, etc.) The parameters a_i, w_{ij}, etc., are called **weights** or **connection strengths**.

By including a larger number of parameters, we are now able to better approximate the optimal test statistic given by the likelihood ratio (4.14), or equivalently by a monotonic function of the likelihood ratio. The problem is now to adjust the parameters so that the resulting $t(\mathbf{x})$ gives an optimal separation between the hypotheses. This is no longer as straightforward as in the linear case, where the parameters could be related to the means and covariances of $\mathbf{x}$. The optimization of the parameters is typically based on minimizing an **error function**, such as

$$\mathcal{E} = E_0[(t - t^{(0)})^2] + E_1[(t - t^{(1)})^2], \tag{4.35}$$

which is analogous to the sum of variances (4.25) minimized for the Fisher discriminant. Here, however, the values $t^{(0)}$ and $t^{(1)}$ represent preassigned **target values** for the hypotheses H_0 and H_1. For a logistic sigmoid activation function for the final layer, the target values are taken to be 0 and 1.

In order to determine the parameters that minimize $\mathcal{E}$, iterative numerical techniques must be used; this is called **network training** or **learning**. In practice, the adjustment of parameters involves replacing the expectation values in (4.35) with the mean values computed from samples of training data, e.g. from a Monte Carlo simulation. Learning algorithms often start with random initial values for the weights and proceed by evaluating the function using some or all of the training data. The weights are then adjusted to minimize $\mathcal{E}$ by one of a variety of methods. A popular procedure is known as **error back-propagation**. A description of these techniques is beyond the scope of this book; more information can be found in [Bis95, Her91, Mül95, Lön92, Pet94].

The choice of the number of layers and number of nodes per layer (the network architecture) depends on the particular problem and on the amount of training data available. For more layers, and hence more parameters, one can achieve a better separation between the two classes. A larger number of parameters will be more difficult to optimize, however, given a finite amount of training data. One can show that a three-layer perceptron is sufficient to provide an arbitrarily good parametrization of any function [Bis95].

4.4.3   Selection of input variables

Up to now we have constructed $t(\mathbf{x})$ using as input variables the entire vector of data $\mathbf{x} = (x_1, \ldots, x_n)$ available for each event. The coefficients $\mathbf{a} = (a_1, \ldots, a_n)$, or the weights in the case of a neural network, will be determined in general from training data, e.g. from a Monte Carlo simulation, and will hence be known only with a finite statistical accuracy. In practice, it is preferable to use only a manageably small subset of the components of $\mathbf{x}$, including only those which contain significant information on the hypotheses in question. It may be that some components contain little or no discriminating power; they can safely be dropped. It may also be that two or more are highly correlated, and thus one does not gain much over using just one of them. By choosing a smaller subset of input variables, one can in general achieve a better determination of the parameters given a finite amount of training data.

One must also keep in mind that the training data may differ in some systematic way from the actual data, e.g. the Monte Carlo simulation will inevitably contain approximations and imperfections. Variables which contribute only marginally to the discrimination between the classes may not in fact be well simulated.

One strategy to choose a subset of the original input variables is to begin by constructing t from the single component of $\mathbf{x}$ which, by itself, gives the best separation between H_0 and H_1. This will be measured for a Fisher discriminant by the value of $J(\mathbf{a})$ (4.22), or in the case of a neural network by the error function $\mathcal{E}$ (4.35). Additional components can then be included one by one such that at each step one achieves the greatest increase in separation. A variation of this procedure would be to begin with the entire set and discard components stepwise such that each step gives the smallest decrease in the separation. Neither procedure guarantees optimality of the result. More on the practical aspects of choosing input variables can be found in [Bis95, Gna88, Her91].

In choosing the input variables, it is important to consider that the original purpose of the statistical test is often to select objects (events, particles, etc.) belonging to a given class in order to study them further. This implies that the properties of these objects are not completely known, otherwise we would not have to carry out the measurement. In deciding which quantities to use as input, one must avoid variables that are correlated with those that are to be studied in a later part of the analysis. This is often a serious constraint, especially since the correlations may not be well understood.

4.5   Goodness-of-fit tests

Frequently one wants to give a measure of how well a given null hypothesis H_0 is compatible with the observed data without specific reference to any alternative hypothesis. This is called a test of the **goodness-of-fit**, and can be done by constructing a test statistic whose value itself reflects the level of agreement between the observed measurements and the predictions of H_0. Procedures for constructing appropriate test statistics will be discussed in Sections 4.7, 6.11 and 7.5. Here we will give a short example to illustrate the main idea.

Suppose one tosses a coin N times and obtains n_h heads and $n_t = N - n_h$ tails. To what extent are n_h and n_t consistent with the hypothesis that the coin is 'fair', i.e. that the probabilities for heads and tails are equal? As a test statistic one can simply use the number of heads n_h, which for a fair coin is assumed to follow a binomial distribution (equation (2.2)) with the parameter $p = 0.5$. That is, the probability to observe heads n_h times is

$$f(n_{\rm h}; N) = \frac{N!}{n_{\rm h}!(N - n_{\rm h})!} \left(\frac{1}{2}\right)^{n_{\rm h}} \left(\frac{1}{2}\right)^{N - n_{\rm h}}.$$ (4.36)

Suppose that $N = 20$ tosses are made and $n_{\rm h} = 17$ heads are observed. Since the expectation value of $n_{\rm h}$ (equation (2.3)) is $E[n_{\rm h}] = Np = 10$, there is evidently a sizable discrepancy between the expected and actually observed outcomes. In order to quantify the significance of the difference one can give the probability of obtaining a result with the same level of discrepancy with the hypothesis or higher. In this case, this is the sum of the probabilities for $n_{\rm h} = 0, 1, 2, 3, 17, 18, 19, 20$. Using equation (4.36) one obtains the probability $P = 0.0026$.

The result of the goodness-of-fit test is thus given by stating the so-called **P-value**, i.e. the probability P, under assumption of the hypothesis in question H_0, of obtaining a result as compatible or less with H_0 than the one actually observed. The P-value is sometimes also called the **observed significance level** or **confidence level**[3] of the test. That is, if we had specified a critical region for the test statistic with a significance level α equal to the P-value obtained, then the value of the statistic would be at the boundary of this region. In a goodness-of-fit test, however, the P-value is a random variable. This is in contrast to the situation in Section 4.1, where the significance level α was a constant specified before carrying out the test.

In the classical approach one stops here, and does not attempt to give a probability for H_0 to be true, since a hypothesis is not treated as a random variable. (The significance level or P-value is, however, often incorrectly interpreted as such a probability.) In Bayesian statistics one would use Bayes' theorem (1.6) to assign a probability to H_0, but this requires giving a prior probability, i.e. the probability that the coin is fair before having seen the outcome of the experiment. In some cases this is a practical approach, in others not. For the present we will remain within the classical framework and simply give the P-value.

The P-value is thus the fraction of times one would obtain data as compatible with H_0 or less so if the experiment (i.e. 20 coin tosses) were repeated many times under similar circumstances. By 'similar circumstances' one means *always with 20 tosses*, or in general with the same number of observations in each experiment. Suppose the experiment had been designed to toss the coin until at least three heads and three tails were observed and then to stop, and in the real experiment this happened to occur after the 20th toss. Assuming such a design, one can show that the probability to stop after the 20th toss or later (i.e. to have an outcome as compatible or less with H_0) is not 0.26% but rather 0.072%, which would seem to lead to a significantly different conclusion about the validity of H_0. Maybe we do not even know how the experimenter decided to toss the coin; we are merely presented with the results afterwards. One way to avoid difficulties with the so-called **optional stopping** problem is simply to interpret 'similar experiments' to always mean experiments with the same number of observations. Although this

---

[3] This is related but not equal to the confidence level of a confidence interval, cf. Section 9.2.

is an arbitrary convention, it allows for a unique interpretation of a reported significance level. For further discussion of this problem see [Ber88, Oha94].

In the example with the coin tosses, the test statistic $t = n_h$ was reasonable since from the symmetry of the problem it was easy to identify the region of values of t that have an equal or lesser degree of compatibility with the hypothesis than the observed value. This is related to the fact that in the case of the coin, the set of all possible alternative hypotheses consists simply of all values of the parameter p not equal to 0.5, and all of these lead to an expected asymmetry between the number of heads and tails.

4.6 The significance of an observed signal

A simple type of goodness-of-fit test is often carried out to judge whether a discrepancy between data and expectation is sufficiently significant to merit a claim for a new discovery. Here one may see evidence for a special type of signal event, the number n_s of which can be treated as a Poisson variable with mean ν_s. In addition to the signal events, however, one will find in general a certain number of background events n_b. Suppose this can also be treated as a Poisson variable with mean ν_b, which we will assume for the moment to be known without error. The total number of events found, $n = n_s + n_b$, is therefore a Poisson variable with mean $\nu = \nu_s + \nu_b$. The probability to observe n events is thus

$$f(n; \nu_s, \nu_b) = \frac{(\nu_s + \nu_b)^n}{n!} e^{-(\nu_s + \nu_b)}. \tag{4.37}$$

Suppose we have carried out the experiment and found n_{obs} events. In order to quantify our degree of confidence in the discovery of a new effect, i.e. $\nu_s \neq 0$, we can compute how likely it is to find n_{obs} events or more from background alone. This is given by

$$P(n \geq n_{obs}) = \sum_{n=n_{obs}}^{\infty} f(n; \nu_s = 0, \nu_b) = 1 - \sum_{n=0}^{n_{obs}-1} f(n; \nu_s = 0, \nu_b)$$

$$= 1 - \sum_{n=0}^{n_{obs}-1} \frac{\nu_b^n}{n!} e^{-\nu_b}. \tag{4.38}$$

For example, if we expect $\nu_b = 0.5$ background events and we observe $n_{obs} = 5$, then the P-value from (4.38) is 1.7×10^{-4}. It should be emphasized that this is not the probability of the hypothesis $\nu_s = 0$. It is rather the probability, under the assumption $\nu_s = 0$, of obtaining as many events as observed or more. Despite this subtlety in its interpretation, the P-value is a useful number to consider when deciding whether a new effect has been found.

Often the result of a measurement is given as the estimated value of a parameter plus or minus one standard deviation (we will return to the question of reporting errors in Chapter 9). Since the standard deviation of a Poisson variable is equal to the square root of its mean, we could take $5 \pm \sqrt{5}$ for an estimate of ν, or after subtracting the background, 4.5 ± 2.2 for our estimate of ν_s. This would be misleading, however, since this is only two standard deviations from zero, and hence gives the impression that ν_s is not very incompatible with zero. As we have seen from the P-value, however, this is not the case. Here we need to ask for the probability that a Poisson variable of mean ν_b will fluctuate up to n_{obs} or higher, not for the probability that a variable with mean n_{obs} will fluctuate down to ν_b or lower. The practice of displaying measured values of Poisson variables with an error bar given by the square root of the observed value unfortunately encourages the incorrect interpretation.

An additional danger is that we have assumed ν_b to be known without error. In the example above, if we had $\nu_b = 0.8$ rather than 0.5, the P-value would increase by almost an order of magnitude to 1.4×10^{-3}. It is therefore important to quantify the systematic uncertainty in the background when evaluating the significance of a new effect. This can be done by giving a range of P-values corresponding to a range of reasonable values of ν_b.

Suppose that in addition to counting the events, we measure for each one a variable x. For a first look at the data one typically constructs a histogram, such as the one shown in Fig. 4.3. The theoretical expectation (dashed histogram) can be normalized such that the value in each bin represents the expected number of entries. If all of the x values are independent, the number of entries in each bin is then a Poisson variable. Given the histogram shown in Fig. 4.3, one would naturally ask if the peak corresponds to a new discovery.

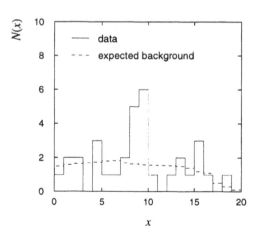

Fig. 4.3 Observed and expected histograms of a variable x. The data show a marginally significant peak.

In principle we can simply apply the procedure above to the number of entries found in any given bin, or any subset of bins. In the two bins with the large peak in Fig. 4.3, there are $n_{obs} = 11$ entries with an expected number of $\nu_b = 3.2$. The probability to observe 11 or more for a Poisson variable with a mean $\nu_b = 3.2$ is 5.0×10^{-4}.

It is usually the case, however, that we do not know a priori where a peak might appear. If a histogram with a large number of bins is constructed from data then naturally some bins will exhibit discrepancies because of expected statistical fluctuations. So in order to give a more meaningful statement of how unexpected the peak is, one could give the probability of observing a discrepancy as improbable as the peak in any of the bins of the histogram. One could argue, however, that this is not sufficient, since we have probably looked at many different histograms as well. But it is in any event more meaningful than reporting the significance level for a specific bin of a selected histogram.

In situations where one is trying to establish the existence of a marginally significant effect, it is important that the event selection and analysis procedures remain fixed once the data have been examined. If they are modified to enhance the significance of the signal (and it is usually impossible to say whether such an enhancement was intentional), then the previous interpretation of the significance level no longer holds. One can, however, modify the analysis procedure and then apply it to a new set of data, and then compute the P-value for the particular bins where the peak was observed in the first data sample. This is of course only possible if additional data are available.

The approach still has drawbacks, however, since the bins outside of the peak region in Fig. 4.3 also depart somewhat from the expected values – some higher, some lower – and this should somehow be taken into account in the evaluation of the overall level of agreement. In addition, the number of entries in the peak bin would change if a different bin size was chosen, which would lead to different values for a test statistic of the type described above. A typical practice is to define the width of the peak region to be at least several times the expected resolution for the variable x.

4.7 Pearson's χ^2 test

In this section we will examine a goodness-of-fit test that can be applied to the distribution of a variable x. As in Fig. 4.3, one begins with a histogram of the observed x values with N bins. Suppose the number of entries in bin i is n_i, and the number of expected entries is ν_i. We would like to construct a statistic which reflects the level of agreement between observed and expected histograms. No doubt the most commonly used goodness-of-fit test is based on Pearson's χ^2 statistic,

$$\chi^2 = \sum_{i=1}^{N} \frac{(n_i - \nu_i)^2}{\nu_i}. \tag{4.39}$$

If the data $\mathbf{n} = (n_1, \ldots, n_N)$ are Poisson distributed with mean values $\boldsymbol{\nu} = (\nu_1, \ldots, \nu_N)$, and if the number of entries in each bin is not too small (e.g. $n_i \geq 5$) then one can show that the statistic (4.39) will follow a χ^2 distribution, equation (2.34), for N degrees of freedom. This holds regardless of the distribution of the variable x; the χ^2 test is therefore said to be **distribution free**. The restriction on the number of entries is equivalent to the requirement that the n_i be approximately Gaussian distributed.

Since the standard deviation of a Poisson variable with mean ν_i is equal to $\sqrt{\nu_i}$, the χ^2 statistic gives the sum of squares of the deviations between observed and expected values, measured in units of the corresponding standard deviations. A larger χ^2 thus corresponds to a larger discrepancy between data and the hypothesis. The P-value or significance level is therefore given by the integral of the χ^2 distribution from the observed χ^2 to infinity,

$$P = \int_{\chi^2}^{\infty} f(z; n_\mathrm{d}) \, dz, \tag{4.40}$$

where here the number of degrees of freedom is $n_\mathrm{d} = N$.[4] Recall that the expectation value of the χ^2 distribution is equal to the number of degrees of freedom. The ratio χ^2/n_d is therefore often quoted as a measure of agreement between data and hypothesis. This does not, however, convey as much information as do χ^2 and n_d separately. For example, the P-value for $\chi^2 = 15$ and $n_\mathrm{d} = 10$ is 0.13. For $\chi^2 = 150$ and 100 degrees of freedom, however, one obtains a P-value of 9.0×10^{-4}.

For equation (4.39) we have assumed that the total number of entries $n_\mathrm{tot} = \sum_{i=1}^{N} n_i$ is itself a Poisson variable with a predicted mean value $\nu_\mathrm{tot} = \sum_{i=1}^{N} \nu_i$. We can, however, regard n_tot as fixed, so that the data n_i are multinomially distributed with probabilities $p_i = \nu_i/n_\mathrm{tot}$. Here one does not test the agreement between the total numbers of expected and observed events, but rather only the distribution of the variable x. One can then construct the χ^2 statistic as

$$\chi^2 = \sum_{i=1}^{N} \frac{(n_i - p_i n_\mathrm{tot})^2}{p_i n_\mathrm{tot}}. \tag{4.41}$$

It can be shown that in the limit where there is a large number of entries in each bin, the statistic (4.41) follows a χ^2 distribution for $N - 1$ degrees of freedom. Here we have assumed that the probabilities p_i are known. In general, if m parameters are estimated from the data, the number of degrees of freedom is reduced by m. We will return to the χ^2 test in Chapter 7 in connection with the method of least squares.

It might be that only a small amount of data is available, so the requirement of at least five entries per bin is not fulfilled. One can still construct the χ^2 statistic, as long as all of the ν_i are greater than zero. It will no longer follow the χ^2 distribution, however, and its distribution will depend on the p.d.f. of the variable x (i.e. the test is no longer distribution free). For the example of Fig. 4.3, for instance, one obtains $\chi^2 = 29.8$ for $n_\mathrm{d} = 20$ degrees of freedom. Here, however, most of the bins have fewer than five entries, and therefore one

---

[4] The cumulative χ^2 distribution, i.e. one minus the integral (4.40), can be computed with the routine **PROB** in [CER97].

cannot regard this as an observation of a χ^2 distributed variable for purposes of computing the P-value.

The correct P-value can be obtained by determining the distribution of the statistic with a Monte Carlo program. This is done by generating Poisson values n_i for each bin based on the mean values ν_i, and then computing and recording the χ^2 value. (Poisson distributed random numbers can be generated with the routine RNPSSN from [CER97].) The distribution resulting from a large number of such experiments is shown in Fig. 4.4 along with the usual χ^2 distribution from equation (2.34). If one were to assume the χ^2 distribution, a P-value of 0.073 would be obtained. The Monte Carlo distribution shows that larger χ^2 values are in fact more probable than this, and gives $P = 0.11$. Note that in this case the χ^2 test is not very sensitive to the presence of the peak, and does not provide significant evidence for rejecting the hypothesis of background with no additional signal.

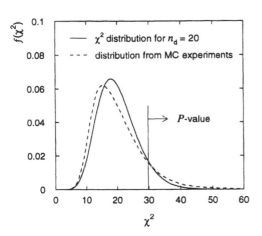

Fig. 4.4 Distribution of the χ^2 statistic from the example of Fig. 4.3 as predicted by the χ^2 distribution and from repeated Monte Carlo experiments.

An ambiguity of the χ^2 test is that one must choose a particular binning. For small data samples, a different choice will in general lead to a different P-value for the test. On the one hand the bins should be sufficiently large that each contains at least about five entries, so that the χ^2 distribution can be used to evaluate the significance level. On the other hand, too large bins throw away information, i.e. the position of the x value within the bin is lost. If the distribution of the χ^2 statistic can be determined by Monte Carlo, then the requirement of a minimum number of entries in each bin can be relaxed and the bin size reduced.

Other tests valid for small data samples are based on the individual x values, i.e. without binning. Examples such as the Kolmogorov–Smirnov and Smirnov–Cramér–von Mises tests are discussed in [Ead71, Fro79].

5

General concepts of parameter estimation

In this chapter some general concepts of parameter estimation are examined which apply to all of the methods discussed in Chapters 6 through 8. In addition, prescriptions for estimating properties of p.d.f.s such as the mean and variance are given.

5.1 Samples, estimators, bias

Consider a random variable x described by a p.d.f. $f(x)$. Here, the sample space is the set of all possible values of x. A set of n independent observations of x is called a **sample** of size n. A new sample space can be defined as the set of all possible values for the n-dimensional vector $\mathbf{x} = (x_1, \ldots, x_n)$. That is, the entire experiment consisting of n measurements is considered to be a single random measurement, which is characterized by n quantities, $x_1, \ldots, x_n$. Since it is assumed that the observations are all independent and that each x_i is described by the same p.d.f. $f(x)$, the joint p.d.f. for the sample $f_{\text{sample}}(x_1, \ldots, x_n)$ is given by

$$f_{\text{sample}}(x_1, \ldots, x_n) = f(x_1)f(x_2) \ldots f(x_n). \qquad (5.1)$$

Although the dimension of the random vector (i.e. the number of measurements) can in practice be very large, the situation is simplified by the fact that the joint p.d.f. for the sample is the product of n p.d.f.s of identical form.

Consider now the situation where one has made n measurements of a random variable x whose p.d.f. $f(x)$ is not known. The central problem of statistics is to infer the properties of $f(x)$ based on the observations $x_1, \ldots, x_n$. Specifically, one would like to construct functions of the x_i to estimate the various properties of the p.d.f. $f(x)$. Often one has a hypothesis for the p.d.f. $f(x; \theta)$ which depends on an unknown parameter θ (or parameters $\boldsymbol{\theta} = (\theta_1, \ldots, \theta_m)$). The goal is then to construct a function of the observed x_i to estimate the parameters.

A function of the observed measurements $x_1, \ldots, x_n$ which contains no unknown parameters is called a **statistic**. In particular, a statistic used to estimate some property of a p.d.f. (e.g. its mean, variance or other parameters) is called an **estimator**. The estimator for a quantity θ is usually written with a hat, $\hat{\theta}$, to distinguish it from the true value θ whose exact value is (and may forever remain) unknown.

If $\hat{\theta}$ converges to θ in the limit of large n, the estimator is said to be **consistent**. Here convergence is meant in the sense of probability, i.e. for any $\epsilon > 0$, one has

$$\lim_{n \to \infty} P(|\hat{\theta} - \theta| > \epsilon) = 0. \tag{5.2}$$

Consistency is usually a minimum requirement for a useful estimator. In the following the limit of large n will be referred to as the 'large sample' or 'asymptotic' limit. In situations where it is necessary to make the distinction, the term estimator will be used to refer to the function of the sample (i.e. its functional form) and an **estimate** will mean the value of the estimator evaluated with a particular sample. The procedure of estimating a parameter's value given the data $x_1, \ldots, x_n$ is called **parameter fitting**.

Since an estimator $\hat{\theta}(x_1, \ldots, x_n)$ is a function of the measured values, it is itself a random variable. That is, if the entire experiment were repeated many times, each time with a new sample $\mathbf{x} = (x_1, \ldots, x_n)$ of size n, the estimator $\hat{\theta}(\mathbf{x})$ would take on different values, being distributed according to some p.d.f. $g(\hat{\theta}; \theta)$, which depends in general on the true value of θ. The probability distribution of a statistic is called a **sampling distribution**. Much of what follows in the next several chapters concerns sampling distributions and their properties, especially expectation value and variance.

The expectation value of an estimator $\hat{\theta}$ with the sampling p.d.f. $g(\hat{\theta}; \theta)$ is

$$E[\hat{\theta}(\mathbf{x})] = \int \hat{\theta} g(\hat{\theta}; \theta) d\hat{\theta}$$

$$= \int \ldots \int \hat{\theta}(\mathbf{x}) f(x_1; \theta) \ldots f(x_n; \theta) dx_1 \ldots dx_n, \tag{5.3}$$

where equation (5.1) has been used for the joint p.d.f. of the sample. Recall that this is the expected mean value of $\hat{\theta}$ from an infinite number of similar experiments, each with a sample of size n. One defines the **bias** of an estimator $\hat{\theta}$ as

$$b = E[\hat{\theta}] - \theta. \tag{5.4}$$

Note that the bias does not depend on the measured values of the sample but rather on the sample size, the functional form of the estimator and on the true (and in general unknown) properties of the p.d.f. $f(x)$, including the true value of θ. A parameter for which the bias is zero independent of the sample size n is said to be unbiased; if the bias vanishes in the limit $n \to \infty$ then it is said to be asymptotically unbiased. Note also that an estimator $\hat{\theta}$ can be biased even if it is consistent. That is, even if $\hat{\theta}$ converges to the true value θ in a single experiment with an infinitely large number of measurements, it does not follow that the average of $\hat{\theta}$ from an infinite number of experiments, each with a finite number of measurements, will converge to the true value. Unbiased estimators are thus particularly valuable if one would like to combine the result with those of other experiments. In most practical cases, the bias is small compared to the statistical error (i.e. the standard deviation) and one does not usually reject using an estimator with a small bias if there are other characteristics (e.g. simplicity) in its favor.

Another measure of the quality of an estimator is the **mean squared error** (MSE), defined as

$$
\begin{aligned}
\text{MSE} \quad &= \quad E[(\hat{\theta} - \theta)^2] = E[(\hat{\theta} - E[\hat{\theta}])^2] + (E[\hat{\theta} - \theta])^2 \\
&= \quad V[\hat{\theta}] + b^2.
\end{aligned}
\tag{5.5}
$$

The MSE is the sum of the variance and the bias squared, and thus can be interpreted as the sum of squares of statistical and systematic errors.

It should be emphasized that classical statistics provides no unique method for constructing estimators. Given an estimator, however, one can say to what extent it has desirable properties, such as small (or zero) bias, small variance, small MSE, etc. We will see in Chapter 6 that there is a certain trade-off between bias and variance. For estimators with a given bias, there is a lower limit to the variance. Often an estimator is deemed 'optimal' if it has zero bias and minimum variance, although other measures of desirability such as the MSE could be considered (cf. Section 11.7). The methods presented in Chapters 6 through 8 will allow us to construct estimators with optimal (or nearly optimal) characteristics in this sense for a wide range of practical cases. Biased estimators are important in inverse problems (unfolding); these are discussed in Chapter 11.

5.2   Estimators for mean, variance, covariance

Suppose one has a sample of size n of a random variable x: $x_1, \ldots, x_n$. It is assumed that x is distributed according to some p.d.f. $f(x)$ which is not known, not even as a parametrization. We would like to construct a function of the x_i to be an estimator for the expectation value of x, μ. One possibility is the arithmetic mean of the x_i, defined by

$$
\bar{x} = \frac{1}{n} \sum_{i=1}^{n} x_i.
\tag{5.6}
$$

The arithmetic mean of the elements of a sample is called the **sample mean**, and is denoted by a bar, e.g. $\bar{x}$. This should not be confused with the expectation value (population mean) of x, denoted by μ or $E[x]$, for which $\bar{x}$ is an estimator.

The first important property of the sample mean is given by the **weak law of large numbers**. This states that if the variance of x exists, then $\bar{x}$ is a consistent estimator for the population mean μ. That is, for $n \to \infty$, $\bar{x}$ converges to μ

in the sense of probability, cf. equation (5.2). A proof can be found in [Bra92]. The condition on the existence of the variance implies, for example, that the law does not hold if x follows the Cauchy distribution (2.40). In that case, in fact, one can show that $\bar{x}$ has the same p.d.f. as x for any sample size. In practice, however, the variances of random variables representing physical quantities are always finite (cf. Section 2.9) and the weak law of large numbers therefore holds.

The expectation value of the sample mean $E[\bar{x}]$ is given by (see equation (5.3))

$$E[\bar{x}] = E\left[\frac{1}{n}\sum_{i=1}^{n} x_i\right] = \frac{1}{n}\sum_{i=1}^{n} E[x_i] = \frac{1}{n}\sum_{i=1}^{n} \mu = \mu, \tag{5.7}$$

since

$$E[x_i] = \int \ldots \int x_i f(x_1) \ldots f(x_n) dx_1 \ldots dx_n = \mu \tag{5.8}$$

for all i. One sees from equation (5.7) that the sample mean $\bar{x}$ is an unbiased estimator for the population mean μ.

The **sample variance** s^2 is defined by

$$s^2 = \frac{1}{n-1}\sum_{i=1}^{n}(x_i - \bar{x})^2 = \frac{n}{n-1}\left(\overline{x^2} - \bar{x}^2\right). \tag{5.9}$$

The expectation value of s^2 can be computed just as was done for the sample mean $\bar{x}$. The factor $1/(n-1)$ is included in the definition of s^2 so that its expectation value comes out equal to σ^2, i.e. so that s^2 is an unbiased estimator for the population variance. If the mean μ is known, then the statistic S^2 defined by

$$S^2 = \frac{1}{n}\sum_{i=1}^{n}(x_i - \mu)^2 = \overline{x^2} - \mu^2 \tag{5.10}$$

is an unbiased estimator of the variance σ^2. In a similar way one can show that the quantity

$$\widehat{V}_{xy} = \frac{1}{n-1}\sum_{i=1}^{n}(x_i - \bar{x})(y_i - \bar{y}) = \frac{n}{n-1}\left(\overline{xy} - \bar{x}\,\bar{y}\right) \tag{5.11}$$

is an unbiased estimator for the covariance V_{xy} of two random variables x and y of unknown mean. This can be normalized by the square root of the estimators for the sample variance to form an estimator r for the correlation coefficient ρ (see equation (1.48); in the following we will often drop the subscripts xy, i.e. here $r = r_{xy}$):

$$r = \frac{\hat{V}_{xy}}{s_x s_y} = \frac{\sum_{i=1}^{n}(x_i - \overline{x})(y_i - \overline{y})}{\left(\sum_{j=1}^{n}(x_j - \overline{x})^2 \cdot \sum_{k=1}^{n}(y_k - \overline{y})^2\right)^{1/2}}$$

$$= \frac{\overline{xy} - \overline{x}\,\overline{y}}{\sqrt{(\overline{x^2} - \overline{x}^2)(\overline{y^2} - \overline{y}^2)}}. \tag{5.12}$$

Given an estimator $\hat{\theta}$ one can compute its variance $V[\hat{\theta}] = E[\hat{\theta}^2] - (E[\hat{\theta}])^2$. Recall that $V[\hat{\theta}]$ (or equivalently its square root $\sigma_{\hat{\theta}}$) is a measure of the variation of $\hat{\theta}$ about its mean in a large number of similar experiments each with sample size n, and as such is often quoted as the statistical error of $\hat{\theta}$. For example, the variance of the sample mean $\overline{x}$ is

$$V[\overline{x}] = E[\overline{x}^2] - (E[\overline{x}])^2 = E\left[\left(\frac{1}{n}\sum_{i=1}^{n} x_i\right)\left(\frac{1}{n}\sum_{j=1}^{n} x_j\right)\right] - \mu^2$$

$$= \frac{1}{n^2}\sum_{i,j=1}^{n} E[x_i x_j] - \mu^2$$

$$= \frac{1}{n^2}\left[(n^2 - n)\mu^2 + n(\mu^2 + \sigma^2)\right] - \mu^2 = \frac{\sigma^2}{n}, \tag{5.13}$$

where σ^2 is the variance of $f(x)$, and we have used the fact that $E[x_i x_j] = \mu^2$ for $i \neq j$ and $E[x_i^2] = \mu^2 + \sigma^2$. This expresses the well-known result that the standard deviation of the mean of n measurements of x is equal to the standard deviation of $f(x)$ itself divided by $\sqrt{n}$.

In a similar way, the variance of the estimator s^2 (5.9) can be computed to be

$$V[s^2] = \frac{1}{n}\left(\mu_4 - \frac{n-3}{n-1}\mu_2^2\right), \tag{5.14}$$

where μ_k is the kth central moment (1.42), e.g. $\mu_2 = \sigma^2$. Using a simple generalization of (5.9), the μ_k can be estimated by

$$m_k = \frac{1}{n-1}\sum_{i=1}^{n}(x_i - \overline{x})^k. \tag{5.15}$$

The expectation value and variance of the estimator of the correlation coefficient r depend on higher moments of the joint p.d.f. $f(x, y)$. For the case of the two-dimensional Gaussian p.d.f. (2.30) they are found to be (see [Mui82] and references therein)

$$E[r] = \rho - \frac{\rho(1 - \rho^2)}{2n} + O(n^{-2}) \qquad (5.16)$$

$$V[r] = \frac{1}{n}(1 - \rho^2)^2 + O(n^{-2}). \qquad (5.17)$$

Although the estimator r given by equation (5.12) is only asymptotically unbiased, it is nevertheless widely used because of its simplicity. Note that although $\widehat{V}_{xy}$, s_x^2 and s_y^2 are unbiased estimators of V_{xy}, σ_x^2 and σ_y^2, the nonlinear function $\widehat{V}_{xy}/(s_x s_y)$ is not an unbiased estimator of $V_{xy}/(\sigma_x \sigma_y)$ (cf. Section 6.2). One should be careful when applying equation (5.17) to evaluate the significance of an observed correlation (see Section 9.5).

6

The method of maximum likelihood

6.1 ML estimators

Consider a random variable x distributed according to a p.d.f. $f(x; \theta)$. Suppose the functional form of $f(x; \theta)$ is known, but the value of at least one parameter θ (or parameters $\theta = (\theta_1, \ldots, \theta_m)$) are not known. That is, $f(x; \theta)$ represents a composite hypothesis for the p.d.f. (cf. Section 4.1). The method of **maximum likelihood** is a technique for estimating the values of the parameters given a finite sample of data. Suppose a measurement of the random variable x has been repeated n times, yielding the values $x_1, \ldots, x_n$. Here, x could also represent a multidimensional random vector, i.e. the outcome of each individual observation could be characterized by several quantities.

Under the assumption of the hypothesis $f(x; \theta)$, including the value of θ, the probability for the first measurement to be in the interval $[x_1, x_1 + dx_1]$ is $f(x_1; \theta)dx_1$. Since the measurements are all assumed to be independent, the probability to have the first one in $[x_1, x_1 + dx_1]$, the second in $[x_2, x_2 + dx_2]$, and so forth is given by

$$\text{probability that } x_i \text{ in } [x_i, x_i + dx_i] \text{ for all } i = \prod_{i=1}^{n} f(x_i; \theta)dx_i. \qquad (6.1)$$

If the hypothesized p.d.f. and parameter values are correct, one expects a high probability for the data that were actually measured. Conversely, a parameter value far away from the true value should yield a low probability for the measurements obtained. Since the dx_i do not depend on the parameters, the same reasoning also applies to the following function L,

$$L(\theta) = \prod_{i=1}^{n} f(x_i; \theta) \qquad (6.2)$$

called the **likelihood function**. Note that this is just the joint p.d.f. for the x_i, although it is treated here as a function of the parameter, θ. The x_i, on the other hand, are treated as fixed (i.e. the experiment is over).

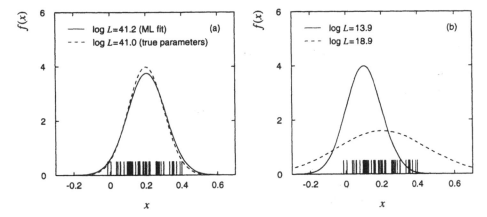

Fig. 6.1 A sample of 50 observations of a Gaussian random variable with mean $\mu = 0.2$ and standard deviation $\sigma = 0.1$. (a) The p.d.f. evaluated with the parameters that maximize the likelihood function and with the true parameters. (b) The p.d.f. evaluated with parameters far from the true values, giving a lower likelihood.

With this motivation one defines the maximum likelihood (ML) estimators for the parameters to be those which maximize the likelihood function. As long as the likelihood function is a differentiable function of the parameters $\theta_1, \ldots, \theta_m$, and the maximum is not at the boundary of the parameter range, the estimators are given by the solutions to the equations,

$$\frac{\partial L}{\partial \theta_i} = 0, \quad i = 1, \ldots, m. \tag{6.3}$$

If more than one local maximum exists, the highest one is taken. As with other types of estimators, they are usually written with hats, $\hat{\boldsymbol{\theta}} = (\hat{\theta}_1, \ldots, \hat{\theta}_m)$, to distinguish them from the true parameters θ_i whose exact values remain unknown.

The general idea of maximum likelihood is illustrated in Fig. 6.1. A sample of 50 measurements (shown as tick marks on the horizontal axis) was generated according to a Gaussian p.d.f. with parameters $\mu = 0.2$, $\sigma = 0.1$. The solid curve in Fig. 6.1(a) was computed using the parameter values for which the likelihood function (and hence also its logarithm) are a maximum: $\hat{\mu} = 0.204$ and $\hat{\sigma} = 0.106$. Also shown as a dashed curve is the p.d.f. using the true parameter values. Because of random fluctuations, the estimates $\hat{\mu}$ and $\hat{\sigma}$ are not exactly equal to the true values μ and σ. The estimators $\hat{\mu}$ and $\hat{\sigma}$ and their variances, which reflect the size of the statistical errors, are derived below in Section 6.3. Figure 6.1(b) shows the p.d.f. for parameters far away from the true values, leading to lower values of the likelihood function.

The motivation for the ML principle presented above does not necessarily guarantee any optimal properties for the resulting estimators. The ML method turns out to have many advantages, among them ease of use and the fact that no binning is necessary. In the following the desirability of ML estimators will be investigated with respect to several criteria, most importantly variance and bias.

6.2   Example of an ML estimator: an exponential distribution

Suppose the proper decay times for unstable particles of a certain type have been measured for n decays, yielding values $t_i, \ldots, t_n$, and suppose one chooses as a hypothesis for the distribution of t an exponential p.d.f. with mean τ:

$$f(t;\tau) = \frac{1}{\tau} e^{-t/\tau}. \tag{6.4}$$

The task here is to estimate the value of the parameter τ. Rather than using the likelihood function as defined in equation (6.2) it is usually more convenient to use its logarithm. Since the logarithm is a monotonically increasing function, the parameter value which maximizes L will also maximize $\log L$. The logarithm has the advantage that the product in L is converted into a sum, and exponentials in f are converted into simple factors. The **log-likelihood function** is thus

$$\log L(\tau) = \sum_{i=1}^{n} \log f(t_i;\tau) = \sum_{i=1}^{n} \left(\log \frac{1}{\tau} - \frac{t_i}{\tau} \right). \tag{6.5}$$

Maximizing $\log L$ with respect to τ gives the ML estimator $\hat{\tau}$,

$$\hat{\tau} = \frac{1}{n} \sum_{i=1}^{n} t_i. \tag{6.6}$$

In this case the ML estimator $\hat{\tau}$ is simply the sample mean of the measured time values. The expectation value of $\hat{\tau}$ is

$$
\begin{aligned}
E[\hat{\tau}(t_1, \ldots, t_n)] &= \int \cdots \int \hat{\tau}(t_1, \ldots, t_n) \, f_{\text{joint}}(t_1, \ldots, t_n; \tau) \, dt_1 \ldots dt_n \\
&= \int \cdots \int \left(\frac{1}{n} \sum_{i=1}^{n} t_i \right) \frac{1}{\tau} e^{-t_1/\tau} \ldots \frac{1}{\tau} e^{-t_n/\tau} dt_1 \ldots dt_n \\
&= \frac{1}{n} \sum_{i=1}^{n} \left(\int t_i \frac{1}{\tau} e^{-t_i/\tau} dt_i \prod_{j \neq i} \int \frac{1}{\tau} e^{-t_j/\tau} dt_j \right) \\
&= \frac{1}{n} \sum_{i=1}^{n} \tau = \tau,
\end{aligned}
\tag{6.7}
$$

so $\hat{\tau}$ is an unbiased estimator for τ. We could have concluded this from the results of Sections 2.4 and 5.2, where it was seen that τ is the expectation value of the exponential p.d.f., and that the sample mean is an unbiased estimator of the expectation value for any p.d.f. (See Section 10.4 for a derivation of the p.d.f. of $\hat{\tau}$.)

As an example consider a sample of 50 Monte Carlo generated decay times t distributed according to an exponential p.d.f. as shown in Fig. 6.2. The values were generated using a true lifetime $\tau = 1.0$. Equation (6.6) gives the ML estimate $\hat{\tau} = 1.062$. The curve shows the exponential p.d.f. evaluated with the ML estimate.

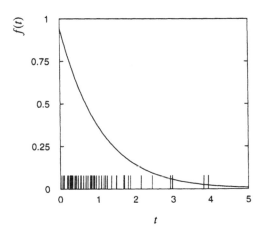

Fig. 6.2 A sample of 50 Monte Carlo generated observations of an exponential random variable t with mean $\tau = 1.0$. The curve is the result of a maximum likelihood fit, giving $\hat{\tau} = 1.062$.

Suppose that one is interested not in the mean lifetime but in the decay constant $\lambda = 1/\tau$. How can we estimate λ? In general, given a function $a(\theta)$ of some parameter θ, one has

$$\frac{\partial L}{\partial \theta} = \frac{\partial L}{\partial a} \frac{\partial a}{\partial \theta} = 0. \tag{6.8}$$

Thus $\partial L/\partial \theta = 0$ implies $\partial L/\partial a = 0$ at $a = a(\theta)$ unless $\partial a/\partial \theta = 0$. As long as this is not the case, one obtains the ML estimator of a function simply by evaluating the function with the original ML estimator, i.e. $\hat{a} = a(\hat{\theta})$. The estimator for the decay constant is thus $\hat{\lambda} = 1/\hat{\tau} = n/\sum_{i=1}^{n} t_i$. The transformation invariance of ML estimators is a convenient property, but an unbiased estimator does not necessarily remain so under transformation. As will be derived in Section 10.4, the expectation value of $\hat{\lambda}$ is

$$E[\hat{\lambda}] = \lambda \frac{n}{n-1} = \frac{1}{\tau} \frac{n}{n-1}, \tag{6.9}$$

so $\hat{\lambda} = 1/\hat{\tau}$ is an unbiased estimator of $1/\tau$ only in the limit of large n, even though $\hat{\tau}$ is an unbiased estimator for τ for any value of n. To summarize, the ML estimator of a function a of a parameter θ is simply $\hat{a} = a(\hat{\theta})$. But if $\hat{\theta}$ is an unbiased estimator of θ ($E[\hat{\theta}] = \theta$) it does not necessarily follow that $a(\hat{\theta})$ is an unbiased estimator of $a(\theta)$. It can be shown, however, that the bias of ML estimators goes to zero in the large sample limit for essentially all practical cases.

(An exception to this rule occurs if the allowed range of the random variable depends on the parameter; see [Ead71] Section 8.3.3.)

6.3   Example of ML estimators: μ and σ^2 of a Gaussian

Suppose one has n measurements of a random variable x assumed to be distributed according to a Gaussian p.d.f. of unknown μ and σ^2. The log-likelihood function is

$$\log L(\mu, \sigma^2) = \sum_{i=1}^{n} \log f(x_i; \mu, \sigma^2) = \sum_{i=1}^{n} \left(\log \frac{1}{\sqrt{2\pi}} + \frac{1}{2} \log \frac{1}{\sigma^2} - \frac{(x_i - \mu)^2}{2\sigma^2} \right).$$
(6.10)

Setting the derivative of $\log L$ with respect to μ equal to zero and solving gives

$$\hat{\mu} = \frac{1}{n} \sum_{i=1}^{n} x_i.$$
(6.11)

Computing the expectation value as done in equation (6.7) gives $E[\hat{\mu}] = \mu$, so $\hat{\mu}$ is unbiased. (As in the case of the mean lifetime estimator $\hat{\tau}$, $\hat{\mu}$ here happens to be a sample mean, so one knows already from Sections 2.5 and 5.2 that it is an unbiased estimator for the mean μ.) Repeating the procedure for σ^2 and using the result for $\hat{\mu}$ gives

$$\widehat{\sigma^2} = \frac{1}{n} \sum_{i=1}^{n} (x_i - \hat{\mu})^2.$$
(6.12)

Computing the expectation value of $\widehat{\sigma^2}$, however, gives

$$E[\widehat{\sigma^2}] = \frac{n-1}{n} \sigma^2.$$
(6.13)

The ML estimator $\widehat{\sigma^2}$ is thus biased, but the bias vanishes in the limit of large n.

Recall, however, from Section 5.1 that the sample variance s^2 is an unbiased estimator for the variance of any p.d.f., so that

$$s^2 = \frac{1}{n-1} \sum_{i=1}^{n} (x_i - \hat{\mu})^2$$
(6.14)

is an unbiased estimator for the parameter σ^2 of the Gaussian. To summarize, equation (6.12) gives the ML estimator for the parameter σ^2, and it has a bias that goes to zero as n approaches infinity. The statistic s^2 from equation (6.14) is not biased (which is good) but it is not the ML estimator.

6.4   Variance of ML estimators: analytic method

Given a set of n measurements of a random variable x and a hypothesis for the
p.d.f. $f(x; \theta)$ we have seen how to estimate its parameters. The next task is to
give some measure of the statistical uncertainty of the estimates. That is, if we
repeated the entire experiment a large number of times (with n measurements
each time) each experiment would give different estimated values for the parameters. How widely spread will they be? One way of summarizing this is with the
variance (or standard deviation) of the estimator.

For certain cases one can compute the variances of the ML estimators analytically. For the example of the exponential distribution with mean τ estimated
by $\hat{\tau} = \frac{1}{n} \sum_{i=1}^{n} t_i$, one has

$$
\begin{aligned}
V[\hat{\tau}] &= E[\hat{\tau}^2] - (E[\hat{\tau}])^2 \\
&= \int \cdots \int \left(\frac{1}{n} \sum_{i=1}^{n} t_i \right)^2 \frac{1}{\tau} e^{-t_1/\tau} \cdots \frac{1}{\tau} e^{-t_n/\tau} dt_1 \ldots dt_n \\
&\quad - \left(\int \cdots \int \left(\frac{1}{n} \sum_{i=1}^{n} t_i \right) \frac{1}{\tau} e^{-t_1/\tau} \cdots \frac{1}{\tau} e^{-t_n/\tau} dt_1 \ldots dt_n \right)^2 \\
&= \frac{\tau^2}{n}.
\end{aligned}
\tag{6.15}
$$

This could have been guessed, since it was seen in Section 5.2 that the variance
of the sample mean is $1/n$ times the variance of the p.d.f. of t (the time of an
individual measurement), for which in this case the variance is τ^2, (Section 2.4)
and the estimator $\hat{\tau}$ happens to be the sample mean.

Remember that the variance of $\hat{\tau}$ computed in equation (6.15) is a function
of the true (and unknown) parameter τ. So what do we report for the statistical
error of the experiment? Because of the transformation invariance of ML estima-
tors (equation (6.8)) we can obtain the ML estimate for the variance $\sigma_{\hat{\tau}}^2 = \tau^2/n$
simply by replacing τ with its own ML estimator $\hat{\tau}$, giving $\widehat{\sigma^2}_{\hat{\tau}} = \hat{\tau}^2/n$, or
similarly for the standard deviation, $\hat{\sigma}_{\hat{\tau}} = \hat{\tau}/\sqrt{n}$.

When an experimenter then reports a result like $\hat{\tau} = 7.82 \pm 0.43$, it is meant
that the estimate (e.g. from ML) is 7.82, and if the experiment were repeated
many times with the same number of measurements per experiment, one would
expect the standard deviation of the distribution of the estimates to be 0.43.
This is one possible interpretation of the 'statistical error' of a fitted parameter,
and is independent of exactly how (according to what p.d.f.) the estimates are
distributed. It is not, however, the standard interpretation in those cases where
the distribution of estimates from many repeated experiments is not Gaussian.
In such cases one usually gives the so-called 68.3% confidence interval, which
will be discussed in Chapter 9. This is the same as plus or minus one standard
deviation if the p.d.f. for the estimator is Gaussian. It can be shown (see e.g.
[Stu91] Section 18.5) that in the large sample limit, ML estimates are in fact

distributed according to a Gaussian p.d.f., so in this case the two procedures lead to the same result.

6.5   Variance of ML estimators: Monte Carlo method

For cases that are too difficult to solve analytically, the distribution of the ML estimates can be investigated with the Monte Carlo method. To do this one must simulate a large number of experiments, compute the ML estimates each time and look at how the resulting values are distributed. For the 'true' parameter in the Monte Carlo program the estimated value from the real experiment can be used. As has been seen in the previous section, the quantity s^2 defined by equation (5.9) is an unbiased estimator for the variance of a p.d.f. Thus one can compute s for the ML estimates obtained from the Monte Carlo experiments and give this as the statistical error of the parameter estimated from the real measurement.

As an example of this technique, consider again the case of the mean lifetime measurement with the exponential distribution (Section 6.2). Using a true lifetime of $\tau = 1.0$, a sample of $n = 50$ measurements gave the ML estimate $\hat{\tau} = 1.062$ (see Fig. 6.2). Regarding the first Monte Carlo experiment as the 'real' one, 1000 further experiments were simulated with 50 measurements each. For these, the true value of the parameter was taken to be $\tau = 1.062$, i.e. the ML estimate of the first experiment.

Figure 6.3 shows a histogram of the resulting ML estimates. The sample mean of the estimates is $\bar{\hat{\tau}} = 1.059$, which is close to the input value, as expected since the ML estimator $\hat{\tau}$ is unbiased. The sample standard deviation from the 1000 experiments is $s = 0.151$. This gives essentially the same error value as what one would obtain from equation (6.15), $\hat{\sigma}_{\hat{\tau}} = \hat{\tau}/\sqrt{n} = 1.062/\sqrt{50} = 0.150$. For the real measurement one would then report (for either method to estimate the error) $\hat{\tau} = 1.06 \pm 0.15$. Note that the distribution is approximately Gaussian in shape. This is a general property of ML estimators for the large sample limit, known as **asymptotic normality**. For a further discussion see, e.g., [Ead71] Chapter 7.

6.6   Variance of ML estimators: the RCF bound

It turns out in many applications to be too difficult to compute the variances analytically, and a Monte Carlo study usually involves a significant amount of work. In such cases one typically uses the **Rao–Cramér–Frechet (RCF) inequality**, also called the **information inequality**, which gives a lower bound on an estimator's variance. This inequality applies to any estimator, not only those constructed from the ML principle. For the case of a single parameter θ the limit is given by

$$V[\hat{\theta}] \geq \left(1 + \frac{\partial b}{\partial \theta}\right)^2 \Big/ E\left[-\frac{\partial^2 \log L}{\partial \theta^2}\right], \qquad (6.16)$$

where b is the bias as defined in equation (5.4) and L is the likelihood function. A proof can be found in [Bra92]. Equation (6.16) is not, in fact, the most general form of the RCF inequality, but the conditions under which the form presented

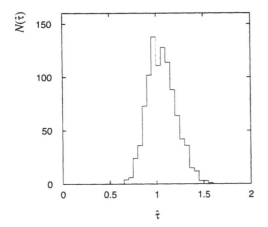

Fig. 6.3 A histogram of the ML estimate $\hat{\tau}$ from 1000 Monte Carlo experiments with 50 observations per experiment. For the Monte Carlo 'true' parameter τ, the result of Fig. 6.2 was used. The sample standard deviation is $s = 0.151$.

here holds are almost always met in practical situations (cf. [Ead71] Section 7.4.5). In the case of equality (i.e. minimum variance) the estimator is said to be **efficient**. It can be shown that if efficient estimators exist for a given problem, the maximum likelihood method will find them. Furthermore it can be shown that ML estimators are always efficient in the large sample limit, except when the extent of the sample space depends on the estimated parameter. In practice, one often assumes efficiency and zero bias. In cases of doubt one should check the results with a Monte Carlo study. The general conditions for efficiency are discussed in, for example, [Ead71] Section 7.4.5, [Stu91] Chapter 18.

For the example of the exponential distribution with mean τ one has from equation (6.5)

$$\frac{\partial^2 \log L}{\partial \tau^2} = \frac{n}{\tau^2} \left(1 - \frac{2}{\tau} \frac{1}{n} \sum_{i=1}^{n} t_i \right) = \frac{n}{\tau^2} \left(1 - \frac{2\hat{\tau}}{\tau} \right) \tag{6.17}$$

and $\partial b / \partial \tau = 0$ since $b = 0$ (see equation (6.7)). Thus the RCF bound for the variance (also called the minimum variance bound, or MVB) of $\hat{\tau}$ is

$$V[\hat{\tau}] \geq \frac{1}{E\left[-\frac{n}{\tau^2}\left(1 - \frac{2\hat{\tau}}{\tau}\right)\right]} = \frac{1}{-\frac{n}{\tau^2}\left(1 - \frac{2E[\hat{\tau}]}{\tau}\right)} = \frac{\tau^2}{n}, \tag{6.18}$$

where we have used equation (6.7) for $E[\hat{\tau}]$. Since τ^2/n is also the variance obtained from the exact calculation (equation (6.15)) we see that equality holds and $\hat{\tau} = \frac{1}{n}\sum_{i=1}^{n} t_i$ is an efficient estimator for the parameter τ.

For the case of more than one parameter, $\boldsymbol{\theta} = (\theta_1, \ldots, \theta_m)$, the corresponding formula for the inverse of the covariance matrix of their estimators $V_{ij} = \text{cov}[\hat{\theta}_i, \hat{\theta}_j]$ is (assuming efficiency and zero bias)

$$(V^{-1})_{ij} = E\left[-\frac{\partial^2 \log L}{\partial \theta_i \partial \theta_j}\right]. \tag{6.19}$$

Equation (6.19) can also be written as

$$
\begin{aligned}
(V^{-1})_{ij} &= \int \cdots \int -\frac{\partial^2}{\partial\theta_i\,\partial\theta_j}\left(\sum_{k=1}^{n}\log f(x_k;\boldsymbol{\theta})\right)\prod_{l=1}^{n}f(x_l;\boldsymbol{\theta})dx_l \\
&= n\cdot\int -f(x;\boldsymbol{\theta})\frac{\partial^2}{\partial\theta_i\,\partial\theta_j}\log f(x;\boldsymbol{\theta})dx, \qquad (6.20)
\end{aligned}
$$

where $f(x;\boldsymbol{\theta})$ is the p.d.f. for the random variable x, for which one has n measurements. That is, the inverse of the RCF bound for the covariance matrix (also called the **Fisher information matrix**, see [Ead71] Section 5.2 and [Bra92]) is proportional to the number of measurements in the sample, n. For $V^{-1}\propto n$ one has $V\propto 1/n$, and thus equation (6.20) expresses the well-known result that statistical errors (i.e. the standard deviations) decrease in proportion to $1/\sqrt{n}$ (at least for efficient estimators).

It turns out to be impractical in many situations to compute the RCF bound analytically, since this requires the expectation value of the second derivative of the log-likelihood function (i.e. an integration over the variable x). In the case of a sufficiently large data sample, one can estimate V^{-1} by evaluating the second derivative with the measured data and the ML estimates $\hat{\boldsymbol{\theta}}$:

$$
(\widehat{V^{-1}})_{ij} = -\frac{\partial^2\log L}{\partial\theta_i\,\partial\theta_j}\bigg|_{\boldsymbol{\theta}=\hat{\boldsymbol{\theta}}}. \qquad (6.21)
$$

For a single parameter θ this reduces to

$$
\widehat{\sigma^2}_{\hat{\theta}} = \left(-1\bigg/\frac{\partial^2\log L}{\partial\theta^2}\right)\bigg|_{\theta=\hat{\theta}}. \qquad (6.22)
$$

This is the usual method for estimating the covariance matrix when the likelihood function is maximized numerically.[1]

6.7   Variance of ML estimators: graphical method

A simple extension of the previously discussed method using the RCF bound leads to a graphical technique for obtaining the variance of ML estimators. Consider the case of a single parameter θ, and expand the log-likelihood function in a Taylor series about the ML estimate $\hat{\theta}$:

$$
\log L(\theta) = \log L(\hat{\theta}) + \left[\frac{\partial\log L}{\partial\theta}\right]_{\theta=\hat{\theta}}(\theta-\hat{\theta}) + \frac{1}{2!}\left[\frac{\partial^2\log L}{\partial\theta^2}\right]_{\theta=\hat{\theta}}(\theta-\hat{\theta})^2 + \cdots .
$$

$$(6.23)$$

---

[1]For example, the routines MIGRAD and HESSE in the program MINUIT [Jam89, CER97] determine numerically the matrix of second derivatives of $\log L$ using finite differences, evaluate it at the ML estimates, and invert to find the covariance matrix.

By definition of $\hat{\theta}$ we know that $\log L(\hat{\theta}) = \log L_{max}$ and that the second term in the expansion is zero. Using equation (6.22) and ignoring higher order terms gives

$$\log L(\theta) = \log L_{max} - \frac{(\theta - \hat{\theta})^2}{2\widehat{\sigma^2}_{\hat{\theta}}}, \tag{6.24}$$

or

$$\log L(\hat{\theta} \pm \hat{\sigma}_{\hat{\theta}}) = \log L_{max} - \frac{1}{2}. \tag{6.25}$$

That is, a change in the parameter θ of one standard deviation from its ML estimate leads to a decrease in the log-likelihood of $1/2$ from its maximum value.

It can be shown that the log-likelihood function becomes a parabola (i.e. the likelihood function becomes a Gaussian curve) in the large sample limit. Even if $\log L$ is not parabolic, one can nevertheless adopt equation (6.25) as the definition of the statistical error. The interpretation of such errors is discussed further in Chapter 9.

As an example of the graphical method for determining the variance of an estimator, consider again the examples of Sections 6.2 and 6.5 with the exponential distribution. Figure 6.4 shows the log-likelihood function $\log L(\tau)$ as a function of the parameter τ for a Monte Carlo experiment consisting of 50 measurements. The standard deviation of $\hat{\tau}$ is estimated by changing τ until $\log L(\tau)$ decreases by $1/2$, giving $\Delta \hat{\tau}_- = 0.137$, $\Delta \hat{\tau}_+ = 0.165$. In this case $\log L(\tau)$ is reasonably close to a parabola and one can approximate $\hat{\sigma}_{\hat{\tau}} \approx \Delta \hat{\tau}_- \approx \Delta \hat{\tau}_+ \approx 0.15$. This leads to approximately the same answer as from the exact standard deviation $\tau/\sqrt{n}$ evaluated with $\tau = \hat{\tau}$. In Chapter 9 the interval $[\hat{\tau} - \Delta \hat{\tau}_-, \hat{\tau} + \Delta \hat{\tau}_+]$ will be reinterpreted as an approximation for the 68.3% **central confidence interval** (cf. Section 9.6).

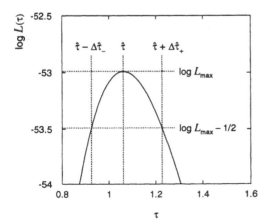

Fig. 6.4 The log-likelihood function $\log L(\tau)$. In the large sample limit, the widths of the intervals $[\hat{\tau} - \Delta \hat{\tau}_-, \hat{\tau}]$ and $[\hat{\tau}, \hat{\tau} + \Delta \hat{\tau}_+]$ correspond to one standard deviation $\hat{\sigma}_{\hat{\tau}}$.

6.8 Example of ML with two parameters

As an example of the maximum likelihood method with two parameters, consider
a particle reaction where each scattering event is characterized by a certain
scattering angle θ (or equivalently $x = \cos\theta$). Suppose a given theory predicts
the angular distribution

$$f(x; \alpha, \beta) = \frac{1 + \alpha x + \beta x^2}{2 + 2\beta/3}. \tag{6.26}$$

(For example, $\alpha = 0$ and $\beta = 1$ correspond to the reaction $e^+ e^- \rightarrow \mu^+ \mu^-$ in lowest
order quantum electrodynamics [Per87].) Note that the denominator $2 + 2\beta/3$ is
necessary for $f(x; \alpha, \beta)$ to be normalized to one for $-1 \leq x \leq 1$.

To make the problem slightly more complicated (and more realistic) assume
that the measurement is only possible in a restricted range, say $x_{min} \leq x \leq x_{max}$.
This requires a recalculation of the normalization constant, giving

$$f(x; \alpha, \beta) = \frac{1 + \alpha x + \beta x^2}{(x_{max} - x_{min}) + \frac{\alpha}{2}(x_{max}^2 - x_{min}^2) + \frac{\beta}{3}(x_{max}^3 - x_{min}^3)}. \tag{6.27}$$

Figure 6.5 shows a histogram of a Monte Carlo experiment where 2000 events
were generated using $\alpha = 0.5$, $\beta = 0.5$, $x_{min} = -0.95$ and $x_{max} = 0.95$. By
numerically maximizing the log-likelihood function one obtains

$$\hat{\alpha} = 0.508 \pm 0.052,$$

$$\tag{6.28}$$

$$\hat{\beta} = 0.47 \pm 0.11,$$

where the statistical errors are the square roots of the variance. These have
been estimated by computing (numerically) the matrix of second derivatives
of the log-likelihood function with respect to the parameters, as described in
Section 6.6, and then inverting to obtain the covariance matrix. From this one
obtains as well an estimate of the covariance $\widehat{\text{cov}}[\hat{\alpha}, \hat{\beta}] = 0.0026$ or equivalently
the correlation coefficient $r = 0.46$. One sees that the estimators $\hat{\alpha}$ and $\hat{\beta}$ are
positively correlated. Note that the histogram itself is not used in the procedure;
the individual values of x are used to compute the likelihood function.

To understand these results more intuitively, it is useful to look at a Monte
Carlo study of 500 similar experiments, all with 2000 events with $\alpha = 0.5$ and
$\beta = 0.5$. A scatter plot of the ML estimates $\hat{\alpha}$ and $\hat{\beta}$ are shown in Fig. 6.6(a).
The density of points corresponds to the joint p.d.f. for $\hat{\alpha}$ and $\hat{\beta}$. Also shown in
Fig. 6.6 (b) and (c) are the normalized projected histograms for $\hat{\alpha}$ and $\hat{\beta}$ sepa-
rately, corresponding to the marginal p.d.f.s, i.e. the distribution of $\hat{\alpha}$ integrated
over all values of $\hat{\beta}$, and vice versa. One sees that the marginal p.d.f.s for $\hat{\alpha}$ and
$\hat{\beta}$ are both approximately Gaussian in shape.

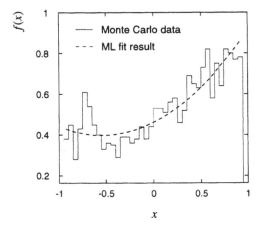

Fig. 6.5 Histogram based on 2000 Monte Carlo generated values distributed according to equation (6.27) with α = 0.5, β = 0.5. Also shown is the result of the ML fit, which gave $\hat{\alpha}$ = 0.508 ± 0.052 and $\hat{\beta}$ = 0.466 ± 0.108. The errors were computed numerically using equation (6.21).

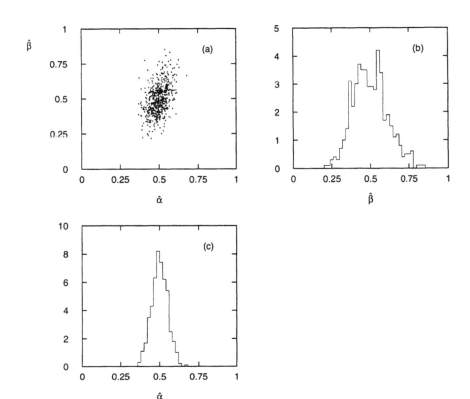

Fig. 6.6 Results of ML fits to 500 Monte Carlo generated data sets. (a) The fitted values of $\hat{\alpha}$ and $\hat{\beta}$. (b) The marginal distribution of $\hat{\beta}$. (c) The marginal distribution of $\hat{\alpha}$.

The sample means, standard deviations, covariance and correlation coefficient (see Section 5.2) from the Monte Carlo experiments are:

$$
\begin{aligned}
\overline{\hat\alpha} &= 0.499 & \overline{\hat\beta} &= 0.498 \\
s_{\hat\alpha} &= 0.051 & s_{\hat\beta} &= 0.111 \\
\widehat{\mathrm{cov}}[\hat\alpha, \hat\beta] &= 0.0024 & r &= 0.42.
\end{aligned}
\tag{6.29}
$$

Note that $\overline{\hat\alpha}$ and $\overline{\hat\beta}$ are in good agreement with the 'true' values put into the Monte Carlo ($\alpha = 0.5$ and $\beta = 0.5$) and the sample (co)variances are close to the values estimated numerically from the RCF bound.

The fact that $\hat\alpha$ and $\hat\beta$ are correlated is seen from the fact that the band of points in the scatter plot is tilted. That is, if one required $\hat\alpha > \alpha$, this would lead to an enhanced probability to also find $\hat\beta > \beta$. In other words, the conditional p.d.f. for $\hat\alpha$ given $\hat\beta > \beta$ is centered at a higher mean value and has a smaller variance than the marginal p.d.f. for $\hat\alpha$.

Figure 6.7 shows the positions of the ML estimates in the parameter space along with a contour corresponding to $\log L = \log L_{\max} - 1/2$.

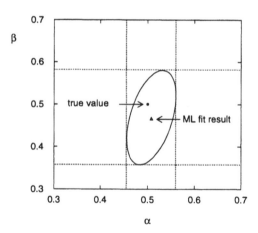

Fig. 6.7 The contour of constant likelihood $\log L = \log L_{\max} - 1/2$ shown with the true values for the parameters (α, β) and the ML estimates $(\hat\alpha, \hat\beta)$. In the large sample limit the tangents to the curve correspond to $\hat\alpha \pm \hat\sigma_{\hat\alpha}$ and $\hat\beta \pm \hat\sigma_{\hat\beta}$.

In the large sample limit, the log-likelihood function takes on the form

$$
\log L(\alpha, \beta) = \log L_{\max}
$$

$$
-\frac{1}{2(1-\rho^2)} \left[\left(\frac{\alpha - \hat\alpha}{\sigma_{\hat\alpha}}\right)^2 + \left(\frac{\beta - \hat\beta}{\sigma_{\hat\beta}}\right)^2 - 2\rho \left(\frac{\alpha - \hat\alpha}{\sigma_{\hat\alpha}}\right) \left(\frac{\beta - \hat\beta}{\sigma_{\hat\beta}}\right) \right], \tag{6.30}
$$

where $\rho = \mathrm{cov}[\hat\alpha, \hat\beta]/(\sigma_{\hat\alpha}\sigma_{\hat\beta})$ is the correlation coefficient for $\hat\alpha$ and $\hat\beta$. The contour of $\log L(\alpha, \beta) = \log L_{\max} - 1/2$ is thus given by

$$\frac{1}{1-\rho^2}\left[\left(\frac{\alpha-\hat{\alpha}}{\sigma_{\hat{\alpha}}}\right)^2+\left(\frac{\beta-\hat{\beta}}{\sigma_{\hat{\beta}}}\right)^2-2\rho\left(\frac{\alpha-\hat{\alpha}}{\sigma_{\hat{\alpha}}}\right)\left(\frac{\beta-\hat{\beta}}{\sigma_{\hat{\beta}}}\right)\right]=1. \qquad (6.31)$$

This is an ellipse centered at the ML estimates $(\hat{\alpha},\hat{\beta})$ and has an angle ϕ with respect to the α axis given by

$$\tan 2\phi = \frac{2\rho\sigma_{\hat{\alpha}}\sigma_{\hat{\beta}}}{\sigma_{\hat{\alpha}}^2-\sigma_{\hat{\beta}}^2}. \qquad (6.32)$$

Note in particular that the tangents to the ellipse are at $\alpha = \hat{\alpha}\pm\sigma_{\hat{\alpha}}, \beta = \hat{\beta}\pm\sigma_{\hat{\beta}}$ (see Fig. 6.7). If the estimators are correlated, then changing a parameter by one standard deviation corresponds in general to a decrease in the log-likelihood of more than $1/2$. If one of the parameters, say β, were known, then the standard deviation of $\hat{\alpha}$ would be somewhat smaller, since this would then be given by a decrease of $1/2$ in $\log L(\alpha)$. Similarly, if additional parameters $(\gamma,\delta,\dots)$ are included in the fit, and if their estimators are correlated with $\hat{\alpha}$, then this will result in an increase in the standard deviation of $\hat{\alpha}$.

6.9 Extended maximum likelihood

Consider a random variable x distributed according to a p.d.f. $f(x;\boldsymbol{\theta})$, with unknown parameters $\boldsymbol{\theta} = (\theta_1,\dots,\theta_m)$, and suppose we have a data sample $x_1,\dots,x_n$. It is often the case that the number of observations n in the sample is itself a Poisson random variable with a mean value ν. The result of the experiment can be defined as the number n and the n values $x_1,\dots,x_n$. The likelihood function is then the product of the Poisson probability to find n, equation (2.9), and the usual likelihood function for the n values of x,

$$L(\nu,\boldsymbol{\theta}) = \frac{\nu^n}{n!}e^{-\nu}\prod_{i=1}^{n}f(x_i;\boldsymbol{\theta}) = \frac{e^{-\nu}}{n!}\prod_{i=1}^{n}\nu\, f(x_i;\boldsymbol{\theta}). \qquad (6.33)$$

This is called the **extended likelihood function**. It is really the usual likelihood function, however, only now with the sample size n defined to be part of the result of the experiment. One can distinguish between two situations of interest, depending on whether the Poisson parameter ν is given as a function of $\boldsymbol{\theta}$ or is treated as an independent parameter.

First assume that ν is given as a function of $\boldsymbol{\theta}$. The extended log-likelihood function is

$$\log L(\boldsymbol{\theta}) \;\; = \;\; n \log \nu(\boldsymbol{\theta}) - \nu(\boldsymbol{\theta}) + \sum_{i=1}^{n} \log f(x_i; \boldsymbol{\theta})$$

$$= \;\; -\nu(\boldsymbol{\theta}) + \sum_{i=1}^{n} \log(\nu(\boldsymbol{\theta}) f(x_i; \boldsymbol{\theta})), \qquad (6.34)$$

where additive terms not depending on the parameters have been dropped. (This is allowed since the estimators depend only on derivatives of $\log L$.) By including the Poisson term, the resulting estimators $\hat{\boldsymbol{\theta}}$ exploit the information from n as well as from the variable x. This leads in general to smaller variances for $\hat{\boldsymbol{\theta}}$ than in the case where only the x values are used.

In a particle scattering reaction, for example, the total cross section as well as the distribution of a variable that characterizes the events, e.g. angles of the outgoing particles, depend on parameters such as particle masses and coupling constants. The statistical errors of the estimated parameters will in general be smaller by including both the information from the cross section as well as from the angular distribution. The total cross section σ is related to the Poisson parameter ν by $\nu = \sigma L \varepsilon$, where L is the integrated luminosity and ε is the probability for an event to be detected (the efficiency). The standard deviations of the estimators correspond to the amount that the estimates would fluctuate if one were to repeat the experiment many times, each time with the same integrated luminosity, and not with the same number of events.

The other situation of interest is where there is no functional relation between ν and $\boldsymbol{\theta}$. Taking the logarithm of (6.33) and setting the derivative with respect to ν equal to zero gives the estimator

$$\hat{\nu} = n, \qquad (6.35)$$

as one would expect. By setting the derivative of $\log L(\nu, \boldsymbol{\theta})$ with respect to the θ_i equal to zero, one obtains the same estimators $\hat{\theta}_i$ as in the usual ML case. So the situation is essentially the same as before, only now a quantity which depends on both n and $\hat{\boldsymbol{\theta}}$ will contain an additional source of statistical fluctuation, since n is regarded as a random variable.

In some problems of this type, however, it can still be helpful to use the extended likelihood function. Often the p.d.f. of a variable x is the superposition of several components,

$$f(x; \boldsymbol{\theta}) \;\; = \;\; \sum_{i=1}^{m} \theta_i f_i(x), \qquad (6.36)$$

and the goal is to estimate the θ_i representing the relative contributions of each component. Suppose that the p.d.f.s $f_i(x)$ are all known. Here the parameters θ_i are not all independent, but rather are constrained to sum to unity. In the usual case without the extended likelihood function, this can be implemented by replacing one of the coefficients, e.g. θ_m, by $1 - \sum_{i=1}^{m-1} \theta_i$, so that the p.d.f.

contains only $m-1$ parameters. One can then construct the likelihood function and from this find estimators for the θ_i.

The problem can be treated in an equivalent but more symmetric way using the extended likelihood function (6.33). Taking the logarithm and dropping terms not depending on the parameters gives

$$\log L(\nu, \boldsymbol{\theta}) = -\nu + \sum_{i=1}^{n} \log \left(\sum_{j=1}^{m} \nu \theta_j f_j(x_i) \right). \tag{6.37}$$

By defining $\mu_i = \theta_i \nu$ as the expected number of events of type i, the log-likelihood function can be written as a function of the m parameters $\boldsymbol{\mu} = (\mu_1, \ldots, \mu_m)$,

$$\log L(\boldsymbol{\mu}) = -\sum_{j=1}^{m} \mu_j + \sum_{i=1}^{n} \log \left(\sum_{j=1}^{m} \mu_j f_j(x_i) \right). \tag{6.38}$$

The parameters $\boldsymbol{\mu}$ are no longer subject to a constraint, as were the components of $\boldsymbol{\theta}$. The total number of events n is viewed as a sum of independent Poisson variables with means μ_i. The estimators $\hat{\mu}_i$ give directly the estimated mean numbers of events of the different types. This is of course equivalent to using the ML estimators $\hat{\theta}_i$ for the fractions along with the estimator $\hat{\nu} = n$ for the single Poisson parameter ν. Now, however, all of the parameters are treated symmetrically. Furthermore, the parameters μ_i are often more closely related to the desired final result, e.g. a production cross section for events of type i.

If the different terms in (6.36) represent different types of events that can contribute to the sample, then one would assume that all of the θ_i are greater than or equal to zero. That is, events of type i can contribute to the sample, but they cannot cause a systematic deficit of events to occur in some region of the distribution of x. One could also consider a function of the form (6.36) where some of the θ_i could be negative, e.g. where some of the $f_i(x)$ represent quantum mechanical interference effects. For now, however, let us consider the case where all of the θ_i are a priori positive or zero. Even in such a case, because of statistical fluctuations in the data it can happen that the likelihood function is maximum when some of the θ_i, and hence the corresponding μ_i, are negative. One must then decide what to report for the estimate.

As an example, consider a data sample consisting of two types of events, e.g. signal and background, where each event is characterized by a continuous variable x. Suppose that for the signal events, $f_s(x)$ is Gaussian distributed, and for the background, $f_b(x)$ is an exponential distribution. The number of signal events n_s is distributed according to a Poisson distribution with mean μ_s, and the number of background events n_b is Poisson distributed with a mean value μ_b.

The p.d.f. for x is thus

$$f(x) = \frac{\mu_s}{\mu_s + \mu_b} f_s(x) + \frac{\mu_b}{\mu_s + \mu_b} f_b(x), \tag{6.39}$$

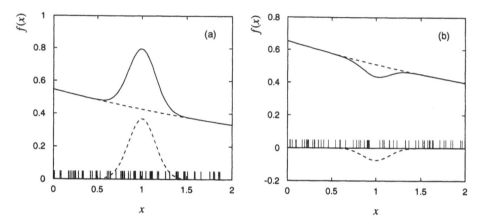

Fig. 6.8 Extended ML fits to two data samples distributed according to a superposition of Gaussian and exponential probability densities. Because of statistical fluctuations in the data, the estimated amplitude of the Gaussian component can turn out positive as in (a), or negative as in (b).

where the forms of the p.d.f.s $f_s(x)$ and $f_b(x)$ are assumed known, and are normalized to unit area within a fixed range, in this example taken to be $0 \leq x \leq 2$. Suppose we are given $n = n_s + n_b$ values of x, and we would like to estimate μ_s. In some cases, the expected background μ_b may be known; in others it may also be determined from the data. In the example here, both μ_s and μ_b are fitted. Figure 6.8 shows two possible data samples generated by Monte Carlo using $\mu_s = 6$ and $\mu_b = 60$, along with the results of the extended ML fit. In Fig. 6.8(a), the estimated signal is $\hat{\mu}_s = 8.7$, and the standard deviation of $\hat{\mu}_s$ estimated from the second derivative of the log-likelihood function is 5.5. Since the standard deviation is comparable to the estimated value, one should not be surprised if such an experiment resulted in a negative estimate. This is in fact what happens in Fig. 6.8(b), which yields $\hat{\mu}_s = -1.8$.

In cases where a negative estimate is physically meaningless, one might choose to take the fitted value if it is positive, but to report zero otherwise. The problem with such an estimator is that it is biased. If one were to perform many similar experiments, some of the ML estimates will be negative and some positive, but the average will converge to the true value. (In principle, ML estimators can still have a bias, but this will be small if the data samples of the individual experiments are sufficiently large, and should be in any event much smaller than the bias introduced by shifting all negative estimates to zero.)

Figure 6.9, for example, shows the estimates $\hat{\mu}_s$ from 200 Monte Carlo experiments of the type above. The average value of $\hat{\mu}_s$ is 6.1, close to the true value of 6. The sample standard deviation of the 200 experiments is 5.3, similar to that estimated above from the log-likelihood function. The standard deviation of the average of the $\hat{\mu}_s$ values is thus $5.3/\sqrt{200} = 0.37$. If the negative estimates are shifted to zero, then the average becomes 6.4. In this example, the bias of

$6.4 - 6 = 0.4$ is small compared to the standard deviation of 5.3, but it could become significant compared to the standard deviation of an average of many experiments.

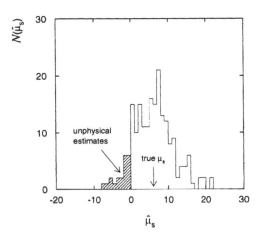

Fig. 6.9 Histogram of estimates $\hat{\mu}_s$ from 200 Monte Carlo experiments based on the true value $\mu_s = 6$. The average value of the estimates is 6.1 ± 0.4. Approximately 10% of the estimates are negative (see text).

Thus if one intends to average a result with that of other experiments, it is important that an unbiased estimate be reported, even if it is in an unphysical region. In addition, one may wish to give an upper limit on the parameter. Methods for this are discussed in Section 9.8.

6.10   Maximum likelihood with binned data

Consider n_{tot} observations of a random variable x distributed according to a p.d.f. $f(x;\boldsymbol{\theta})$ for which we would like to estimate the unknown parameter $\boldsymbol{\theta} = (\theta_1, \ldots, \theta_m)$. For very large data samples, the log-likelihood function becomes difficult to compute since one must sum $\log f(x_i; \boldsymbol{\theta})$ for each value x_i. In such cases, instead of recording the value of each measurement one usually makes a histogram, yielding a certain number of entries $\mathbf{n} = (n_1, \ldots, n_N)$ in N bins. The expectation values $\boldsymbol{\nu} = (\nu_1, \ldots, \nu_N)$ of the numbers of entries are given by

$$\nu_i(\boldsymbol{\theta}) = n_{\text{tot}} \int_{x_i^{\min}}^{x_i^{\max}} f(x;\boldsymbol{\theta})dx, \tag{6.40}$$

where $x_i^{\min}$ and $x_i^{\max}$ are the bin limits. One can regard the histogram as a single measurement of an N-dimensional random vector for which the joint p.d.f. is given by a multinomial distribution, equation (2.6),

$$f_{\text{joint}}(\mathbf{n}; \boldsymbol{\nu}) = \frac{n_{\text{tot}}!}{n_1! \ldots n_N!} \left(\frac{\nu_1}{n_{\text{tot}}}\right)^{n_1} \ldots \left(\frac{\nu_N}{n_{\text{tot}}}\right)^{n_N}. \tag{6.41}$$

The probability to be in bin i has been expressed as the expectation value ν_i divided by the total number of entries n_{tot}. Taking the logarithm of the joint p.d.f. gives the log-likelihood function,

$$\log L(\boldsymbol{\theta}) = \sum_{i=1}^{N} n_i \log \nu_i(\boldsymbol{\theta}), \tag{6.42}$$

where additive terms not depending on the parameters have been dropped. The estimators $\hat{\boldsymbol{\theta}}$ are found by maximizing $\log L$ by whatever means available, e.g. numerically. In the limit that the bin size is very small (i.e. N very large) the likelihood function becomes the same as that of the ML method without binning (equation (6.2)). Thus the binned ML technique does not encounter any difficulties if some of the bins have few or no entries. This is in contrast to an alternative technique using the method of least squares discussed in Section 7.5.

As an example consider again the sample of 50 measured particle decay times that we examined in Section 6.2, for which the ML result without binning is shown in Fig. 6.2. Figure 6.10 shows the same sample displayed as a histogram with a bin width of $\Delta t = 0.5$. Also shown is the fit result obtained from maximizing the log-likelihood function based on equation (6.42). The result is $\hat{\tau} = 1.067$, in good agreement with the unbinned result of $\hat{\tau} = 1.062$. Estimating the standard deviation from the curvature of the log-likelihood at its maximum (equation (6.22)) results in $\hat{\sigma}_{\hat{\tau}} = 0.171$, slightly larger than that obtained without binning.

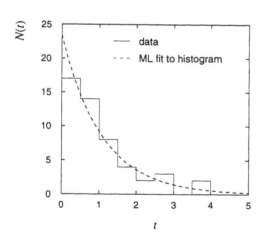

Fig. 6.10 Histogram of the data sample of 50 particle decay times from Section 6.2 with the ML fit result.

As discussed in Section 6.9, in many problems one may want to regard the total number of entries n_{tot} as a random variable from a Poisson distribution with mean ν_{tot}. That is, the measurement is defined to consist of first determining n_{tot} from a Poisson distribution and then distributing n_{tot} observations of x in a histogram with N bins, giving $\mathbf{n} = (n_1, \ldots, n_N)$. The joint p.d.f. for n_{tot} and $n_1, \ldots, n_N$ is the product of a Poisson distribution and a multinomial distribution,

$$f_{\text{joint}}(\mathbf{n}; \boldsymbol{\nu}) = \frac{\nu_{\text{tot}}^{n_{\text{tot}}} e^{-\nu_{\text{tot}}}}{n_{\text{tot}}!} \frac{n_{\text{tot}}!}{n_1! \dots n_N!} \left(\frac{\nu_1}{\nu_{\text{tot}}}\right)^{n_1} \dots \left(\frac{\nu_N}{\nu_{\text{tot}}}\right)^{n_N}, \tag{6.43}$$

where one has $\nu_{\text{tot}} = \sum_{i=1}^{N} \nu_i$ and $n_{\text{tot}} = \sum_{i=1}^{N} n_i$. Using these in equation (6.43) gives

$$f_{\text{joint}}(\mathbf{n}; \boldsymbol{\nu}) = \prod_{i=1}^{N} \frac{\nu_i^{n_i}}{n_i!} e^{-\nu_i}, \tag{6.44}$$

where the expected number of entries in each bin ν_i now depends on the parameters $\boldsymbol{\theta}$ and ν_{tot},

$$\nu_i(\nu_{\text{tot}}, \boldsymbol{\theta}) = \nu_{\text{tot}} \int_{x_i^{\min}}^{x_i^{\max}} f(x; \boldsymbol{\theta}) dx. \tag{6.45}$$

From the joint p.d.f. (6.44) one sees that the problem is equivalent to treating the number of entries in each bin as an independent Poisson random variable n_i with mean value ν_i. Taking the logarithm of the joint p.d.f. and dropping terms that do not depend on the parameters gives

$$\log L(\nu_{\text{tot}}, \boldsymbol{\theta}) = -\nu_{\text{tot}} + \sum_{i=1}^{N} n_i \log \nu_i(\nu_{\text{tot}}, \boldsymbol{\theta}). \tag{6.46}$$

This is the extended log-likelihood function, cf. equations (6.33), (6.37), now for the case of binned data.

The previously discussed considerations on the dependence between ν_{tot} and the other parameters $\boldsymbol{\theta}$ apply in the same way here. That is, if there is no functional relation between ν_{tot} and $\boldsymbol{\theta}$, then one obtains $\hat{\nu}_{\text{tot}} = n_{\text{tot}}$, and the estimates $\hat{\boldsymbol{\theta}}$ come out the same as when the Poisson term for n_{tot} is not included. If ν_{tot} is given as a function of $\boldsymbol{\theta}$, then the variances of the estimators $\hat{\boldsymbol{\theta}}$ are in general reduced by including the information from n_{tot}.

6.11 Testing goodness-of-fit with maximum likelihood

While the principle of maximum likelihood provides a method to estimate parameters, it does not directly suggest a method of testing goodness-of-fit. One possibility is to use the value of the likelihood function at its maximum, $L_{\max}$, as a goodness-of-fit statistic. This is not so simple, however, since one does not know a priori how $L_{\max}$ is distributed assuming that the form of the p.d.f. is correct.

The p.d.f. of $L_{\max}$ can be determined by means of a Monte Carlo study. For the 'true' Monte Carlo parameters used to generate the data, the ML estimates from the real experiment can be used. This was done for the example of the scattering experiment discussed in Section 6.8, and the distribution of $\log L_{\max}$ is shown in Fig. 6.11. The original example (the data set shown in Fig. 6.5) gave $\log L_{\max} = 2436.4$. From this one can compute an observed significance level

(P-value), as described in Section 4.5, as a measure of the goodness-of-fit. For the example here one obtains $P = 0.63$, and so there would not be any evidence against the form of the p.d.f. used in the fit.

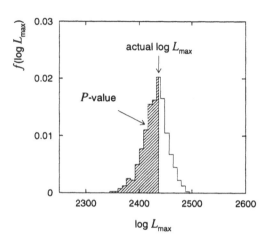

Fig. 6.11 Normalized histogram of the values of the maximized log-likelihood function $\log L_{\max}$ from 500 Monte Carlo experiments. The vertical line shows the value of $\log L_{\max}$ obtained using the data shown in Fig. 6.5 (see text).

Another approach is to construct a histogram $\mathbf{n} = (n_1, \ldots, n_N)$ with N bins from the n_{tot} measured values. The estimates of the mean values $\boldsymbol{\nu} = (\nu_1, \ldots, \nu_N)$ from the ML fit are

$$\hat{\nu}_i = n_{\text{tot}} \int_{x_i^{\min}}^{x_i^{\max}} f(x; \hat{\boldsymbol{\theta}}) dx, \qquad (6.47)$$

where the parameters $\boldsymbol{\theta}$ are evaluated with their ML estimates. This can of course be computed even if the ML fit was done without binning. The histogram offers the advantage that it can be displayed graphically, and as a first step, a visual comparison of the data and fit result can be made. At a more quantitative level, the data $\mathbf{n}$ and the estimates $\hat{\boldsymbol{\nu}}$ (or other hypothesized values $\boldsymbol{\nu}$) can be used to construct a goodness-of-fit statistic.

An example of such a statistic is based on the likelihood function from the binned ML fit in Section 6.10. This is given by the multinomial p.d.f. (6.41) for fixed sample size n_{tot}, or by the product of Poisson probabilities (6.44) if n_{tot} is treated as a Poisson variable with mean ν_{tot}. Consider first the ratio

$$\lambda = \frac{L(\mathbf{n}|\boldsymbol{\nu})}{L(\mathbf{n}|\mathbf{n})} = \frac{f_{\text{joint}}(\mathbf{n}; \boldsymbol{\nu})}{f_{\text{joint}}(\mathbf{n}; \mathbf{n})}, \qquad (6.48)$$

where here the likelihood function $L(\mathbf{n}|\boldsymbol{\nu}) = f_{\text{joint}}(\mathbf{n}; \boldsymbol{\nu})$ is written to emphasize the dependence on both the data $\mathbf{n}$ and the parameters $\boldsymbol{\nu}$. That is, in the denominator of (6.48) the ν_i are set equal to the data values n_i. For multinomially distributed data this becomes

$$\lambda_{\mathrm{M}} = \prod_{i=1}^{N} \left(\frac{\nu_i}{n_i}\right)^{n_i}, \qquad (6.49)$$

and for Poisson distributed data one obtains

$$\lambda_{\mathrm{P}} = e^{n_{\mathrm{tot}} - \nu_{\mathrm{tot}}} \prod_{i=1}^{N} \left(\frac{\nu_i}{n_i}\right)^{n_i}. \qquad (6.50)$$

If m parameters have been estimated from the data, then ν can be replaced by the estimates $\hat{\nu}$. If the hypothesis is correct, then in the large sample limit with multinomially distributed data, the statistic

$$\chi_{\mathrm{M}}^2 = -2\log\lambda_{\mathrm{M}} = 2\sum_{i=1}^{N} n_i \log\frac{n_i}{\nu_i}, \qquad (6.51)$$

follows a χ^2 distribution for $N - m - 1$ degrees of freedom (see [Bak84, Ead71] and references therein). For Poisson distributed data, the statistic

$$\chi_{\mathrm{P}}^2 = -2\log\lambda_{\mathrm{P}} = 2\sum_{i=1}^{N} \left(n_i \log\frac{n_i}{\hat{\nu}_i} + \hat{\nu}_i - n_i\right) \qquad (6.52)$$

follows a χ^2 distribution for $N - m$ degrees of freedom. These quantities appear not to be defined if any n_i are equal to zero, but in such a case the factor $n_i^{n_i}$ in λ is taken to be unity, and the corresponding terms do not contribute in (6.51) and (6.52).

The quantity $\lambda(\nu) = L(\mathbf{n}|\nu)/L(\mathbf{n}|\mathbf{n})$ only differs from the likelihood function by the factor $L(\mathbf{n}|\mathbf{n})$, which does not depend on the parameters. The parameters that maximize $\lambda(\nu)$ are therefore equal to the ML estimators. One can thus use $\lambda(\nu)$ both for constructing estimators as well as for testing goodness-of-fit.

Alternatively, one can use one of the χ^2 statistics from Section 4.7. If n_{tot} is treated as a Poisson variable, one has

$$\chi^2 = \sum_{i=1}^{N} \frac{(n_i - \hat{\nu}_i)^2}{\hat{\nu}_i} \qquad (6.53)$$

from equation (4.39), or if n_{tot} is fixed, then one can use

$$\chi^2 = \sum_{i=1}^{N} \frac{(n_i - \hat{p}_i n_{\mathrm{tot}})^2}{\hat{p}_i n_{\mathrm{tot}}} \qquad (6.54)$$

from equation (4.41). Here $\hat{p}_i = \hat{\nu}_i/\hat{\nu}_{\mathrm{tot}}$ is the estimated probability for a measurement to be found in bin i. In the large sample limit these follow χ^2 distributions with the number of degrees of freedom equal to $N - m$ for (6.53) and $N - m - 1$ for (6.54).

For finite data samples, none of the statistics given above follow exactly the χ^2 distribution. If the histogram contains bins with, for example, $n_i < 5$, a Monte Carlo study can be carried out to determine the true p.d.f., which can then be used to obtain a P-value.

6.12   Combining measurements with maximum likelihood

Consider an experiment in which one has n measured values of a random variable x, for which the p.d.f. $f_x(x; \theta)$ depends on an unknown parameter θ. Suppose in another experiment one has m measured values of a different random variable y, whose p.d.f. $f_y(y; \theta)$ depends on the same parameter θ. For example, x could be the invariant mass of electron–positron pairs produced in proton–antiproton collisions, and y could be the invariant mass of muon pairs. Both distributions have peaks at around the mass M_Z of the Z^0 boson, and so both p.d.f.s contain M_Z as a parameter. One then wishes to combine the two experiments in order to obtain the best estimate of M_Z.

The two experiments together can be interpreted as a single measurement of a vector containing n values of x and m values of y. The likelihood function is therefore

$$L(\theta) = \prod_{i=1}^{n} f_x(x_i; \theta) \cdot \prod_{j=1}^{m} f_y(y_j; \theta) = L_x(\theta) \cdot L_y(\theta), \tag{6.55}$$

or equivalently its logarithm is given by the sum $\log L(\theta) = \log L_x(\theta) + \log L_y(\theta)$.

Thus as long as the likelihood functions of the experiments are available, the full likelihood function can be constructed and the ML estimator for θ based on both experiments can be determined. This technique includes of course the special case where x and y are the same random variable, and the samples $x_1, \ldots, x_n$ and $y_1, \ldots, y_m$ simply represent two different subsamples of the data.

More frequently one does not report the likelihood functions themselves, but rather only estimates of the parameters. Suppose the two experiments based on measurements of x and y give estimators $\hat{\theta}_x$ and $\hat{\theta}_y$ for the parameter θ, which themselves are random variables distributed according to the p.d.f.s $g_x(\hat{\theta}_x; \theta)$ and $g_y(\hat{\theta}_y; \theta)$. The two estimators can be regarded as the outcome of a single experiment yielding the two-dimensional vector $(\hat{\theta}_x, \hat{\theta}_y)$. As long as $\hat{\theta}_x$ and $\hat{\theta}_y$ are independent, the log-likelihood function is given by the sum

$$\log L(\theta) = \log g_x(\hat{\theta}_x; \theta) + \log g_y(\hat{\theta}_y; \theta). \tag{6.56}$$

For large data samples the p.d.f.s g_x and g_y can be assumed to be Gaussian, and one reports the estimated standard deviations $\hat{\sigma}_{\hat{\theta}_x}$ and $\hat{\sigma}_{\hat{\theta}_y}$ as the errors on $\hat{\theta}_x$ and $\hat{\theta}_y$. As will be seen in Chapter 7, the problem is then equivalent to the method of least squares, and the combined estimate for θ is given by the weighted average

$$\hat{\theta} = \frac{\hat{\theta}_x / \hat{\sigma}_{\hat{\theta}_x}^2 + \hat{\theta}_y / \hat{\sigma}_{\hat{\theta}_y}^2}{1 / \hat{\sigma}_{\hat{\theta}_x}^2 + 1 / \hat{\sigma}_{\hat{\theta}_y}^2}, \tag{6.57}$$

with the estimated variance

$$\hat{V}[\hat{\theta}] = \frac{1}{1 / \hat{\sigma}_{\hat{\theta}_x}^2 + 1 / \hat{\sigma}_{\hat{\theta}_y}^2}. \tag{6.58}$$

This technique can clearly be generalized to combine any number of measurements.

6.13 Relationship between ML and Bayesian estimators

It is instructive to compare the method of maximum likelihood to parameter estimation in Bayesian statistics, where uncertainty is quantified by means of subjective probability (cf. Section 1.2). Here, both the result of a measurement x and a parameter θ are treated as random variables. One's knowledge about θ is summarized by a probability density, which expresses the degree of belief for the parameter to take on a given value.

Consider again n observations of a random variable x, assumed to be distributed according to some p.d.f. $f(x; \theta)$, which depends on an unknown parameter θ. (The Bayesian approach can easily be generalized to several parameters $\theta = (\theta_1, \ldots, \theta_m)$. For simplicity we will consider here only a single parameter.) Recall that the likelihood function is the joint p.d.f. for the data $\mathbf{x} = (x_1, \ldots, x_n)$ for a given value of θ, and thus can be written

$$L(\mathbf{x}|\theta) = f_{\text{joint}}(\mathbf{x}|\theta) = \prod_{i=1}^{n} f(x_i; \theta). \tag{6.59}$$

What we would like is the conditional p.d.f. for θ given the data $p(\theta|\mathbf{x})$. This is obtained from the likelihood via Bayes' theorem, equation (1.26),

$$p(\theta|\mathbf{x}) = \frac{L(\mathbf{x}|\theta)\, \pi(\theta)}{\int L(\mathbf{x}|\theta')\, \pi(\theta')d\theta'}. \tag{6.60}$$

Here $\pi(\theta)$ is the **prior probability density** for θ, reflecting the state of knowledge of θ before consideration of the data, and $p(\theta|\mathbf{x})$ is called the **posterior probability density** for θ given the data $\mathbf{x}$.

In Bayesian statistics, all of our knowledge about θ is contained in $p(\theta|\mathbf{x})$. Since it is rarely practical to report the entire p.d.f., especially when θ is multidimensional, an appropriate way of summarizing it must be found. The first step in this direction is an estimator, which is often taken to be the value of θ at which $p(\theta|\mathbf{x})$ is a maximum (i.e. the posterior mode). If the prior p.d.f. $\pi(\theta)$ is taken to be a constant, then $p(\theta|\mathbf{x})$ is proportional to the likelihood function $L(\mathbf{x}|\theta)$, and the Bayesian and ML estimators coincide. The ML estimator can thus be regarded as a special case of a Bayesian estimator, based on a uniform prior pd.f. We should therefore take a closer look at what a uniform $\pi(\theta)$ implies.

The Bayesian approach expressed by equation (6.60) gives a method for updating one's state of knowledge in light of newly acquired data. It is necessary to specify, however, what the state of knowledge was before the measurement was carried out. If nothing is known previously, one may assume that a priori all values of θ are equally likely. This assumption is sometimes called **Bayes' postulate**, expressed here by $\pi(\theta) = $ constant. If the range of θ is infinite, then a constant $\pi(\theta)$ cannot be normalized, and is called an **improper prior**. This is usually not, in fact, a problem since $\pi(\theta)$ always appears multiplied by the likelihood function, resulting in a normalizable posterior p.d.f. For some improper prior densities this may not always be the case; cf. equation (9.45) in Chapter 9.

A more troublesome difficulty with constant prior densities arises when one considers a transformation of parameters. Consider, for example, a continuous parameter θ defined in the interval $[0, 10]$. Using Bayes' postulate, one would use the prior p.d.f. $\pi_\theta(\theta) = 0.1$ in equation (6.60) to obtain the posterior density $p_\theta(\theta|\mathbf{x})$. Another experimenter, however, could decide that some nonlinear function $a(\theta)$ was more appropriate as the parameter. By transformation of variables, one could find the corresponding density $p_a(a|\mathbf{x}) = p_\theta(\theta|\mathbf{x})|d\theta/da|$. Alternatively, one could express the likelihood function directly in terms of a, and assume that the prior density $\pi_a(a)$ is constant. For example, if $a = \theta^2$, then $\pi_a(a) = 0.01$ in the interval $[0, 100]$. Using this in equation (6.60), however, would lead to a posterior density in general different from the $p_a(a|\mathbf{x})$ obtained by transformation of variables. That is, complete ignorance about θ ($\pi_\theta(\theta) = $ constant) implies a nonuniform prior density for a nonlinear function of θ ($\pi_a(a) \neq $ constant).

But if $\pi_a(a)$ is not constant, then the mode of the posterior $p_a(a|\mathbf{x})$ will not occur at the same place as the maximum of the likelihood function $L_a(\mathbf{x}|a) = f_{\text{joint}}(\mathbf{x}|a)$. That is, the Bayesian estimator is not in general invariant under a transformation of the parameter. The ML estimator is, however, invariant under parameter transformation, as noted in Section 6.2. That is, the value of a that maximizes $L_a(\mathbf{x}|a)$ is simply $a(\hat{\theta})$, where $\hat{\theta}$ is the value of θ that maximizes $L_\theta(\mathbf{x}|\theta)$.

7

The method of least squares

7.1 Connection with maximum likelihood

In many situations a measured value y can be regarded as a Gaussian random variable centered about the quantity's true value λ. This follows from the central limit theorem as long as the total error (i.e. deviation from the true value) is the sum of a large number of small contributions.

Consider now a set of N independent Gaussian random variables y_i, $i = 1, \ldots, N$, each related to another variable x_i, which is assumed to be known without error. For example, one may have N measurements of a temperature $T(x_i)$ at different positions x_i. Assume that each value y_i has a different unknown mean, λ_i, and a different but known variance, σ_i^2. The N measurements of y_i can be equivalently regarded as a single measurement of an N-dimensional random vector, for which the joint p.d.f. is the product of N Gaussians,

$$g(y_1, \ldots, y_N; \lambda_1, \ldots, \lambda_N, \sigma_1^2, \ldots, \sigma_N^2) = \prod_{i=1}^{N} \frac{1}{\sqrt{2\pi\sigma_i^2}} \exp\left(\frac{-(y_i - \lambda_i)^2}{2\sigma_i^2}\right). \quad (7.1)$$

Suppose further that the true value is given as a function of x, $\lambda = \lambda(x; \boldsymbol{\theta})$, which depends on unknown parameters $\boldsymbol{\theta} = (\theta_1, \ldots, \theta_m)$. The aim of the method of least squares is to estimate the parameters $\boldsymbol{\theta}$. In addition, the method allows for a simple evaluation of the goodness-of-fit of the hypothesized function $\lambda(x; \boldsymbol{\theta})$. The basic ingredients of the problem are illustrated in Fig. 7.1.

Taking the logarithm of the joint p.d.f. and dropping additive terms that do not depend on the parameters gives the log-likelihood function,

$$\log L(\boldsymbol{\theta}) = -\frac{1}{2} \sum_{i=1}^{N} \frac{(y_i - \lambda(x_i; \boldsymbol{\theta}))^2}{\sigma_i^2}. \quad (7.2)$$

This is maximized by finding the values of the parameters $\boldsymbol{\theta}$ that minimize the quantity

$$\chi^2(\boldsymbol{\theta}) = \sum_{i=1}^{N} \frac{(y_i - \lambda(x_i; \boldsymbol{\theta}))^2}{\sigma_i^2}, \quad (7.3)$$

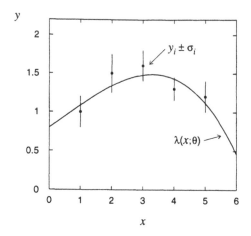

Fig. 7.1 Ingredients of the least squares problem: N values $y_1, \ldots, y_N$ are measured with errors $\sigma_1, \ldots, \sigma_N$ at the values of x given without error by $x_1, \ldots, x_N$. The true value λ_i of y_i is assumed to be given by a function $\lambda_i = \lambda(x_i; \boldsymbol{\theta})$. The value of $\boldsymbol{\theta}$ is adjusted to minimize the value of χ^2 given by equation (7.3).

namely the quadratic sum of the differences between measured and hypothesized values, weighted by the inverse of the variances. This is the basis of the **method of least squares** (LS), and is used to define the procedure even in cases where the individual measurements y_i are not Gaussian, but as long as they are independent.

If the measurements are not independent but described by an N-dimensional Gaussian p.d.f. with known covariance matrix V but unknown mean values, the corresponding log-likelihood function is obtained from the logarithm of the joint p.d.f. given by equation (2.28),

$$\log L(\boldsymbol{\theta}) = -\frac{1}{2} \sum_{i,j=1}^{N} (y_i - \lambda(x_i; \boldsymbol{\theta}))(V^{-1})_{ij}(y_j - \lambda(x_j; \boldsymbol{\theta})), \qquad (7.4)$$

where additive terms not depending on the parameters have been dropped. This is maximized by minimizing the quantity

$$\chi^2(\boldsymbol{\theta}) = \sum_{i,j=1}^{N} (y_i - \lambda(x_i; \boldsymbol{\theta}))(V^{-1})_{ij}(y_j - \lambda(x_j; \boldsymbol{\theta})), \qquad (7.5)$$

which reduces to equation (7.3) if the covariance matrix (and hence its inverse) are diagonal.

The parameters that minimize the χ^2 are called the LS estimators, $\hat{\theta}_1, \ldots, \hat{\theta}_m$. As will be discussed in Section 7.5, the resulting minimum χ^2 follows under certain circumstances the χ^2 distribution, as defined in Section 2.7. Because of this the quantity defined by equations (7.3) or (7.5) is often called χ^2, even in more general circumstances where its minimum value is not distributed according to the χ^2 p.d.f.

7.2   Linear least-squares fit

Although one can carry out the least squares procedure for any function $\lambda(x; \boldsymbol{\theta})$, the resulting χ^2 value and LS estimators have particularly desirable properties for the case where $\lambda(x; \boldsymbol{\theta})$ is a linear function of the parameters $\boldsymbol{\theta}$,

$$\lambda(x; \boldsymbol{\theta}) = \sum_{j=1}^{m} a_j(x)\theta_j, \tag{7.6}$$

where the $a_j(x)$ are any linearly independent functions of x. (What is required is that λ is linear in the parameters θ_j. The $a_j(x)$ are not in general linear in x, but are just linearly independent from each other, i.e. one cannot be expressed as a linear combination of the others.) For this case, the estimators and their variances can be found analytically, although depending on the tools available one may still prefer to maximize χ^2 numerically with a computer. Furthermore, the estimators have zero bias and minimum variance. This follows from the Gauss–Markov theorem (see e.g. [Stu91]) and holds regardless of the number of measurements N, and the p.d.f.s of the individual measurements.

The value of the function $\lambda(x; \boldsymbol{\theta})$ at x_i can be written

$$\lambda(x_i; \boldsymbol{\theta}) = \sum_{j=1}^{m} a_j(x_i)\theta_j = \sum_{j=1}^{m} A_{ij}\theta_j \tag{7.7}$$

where $A_{ij} = a_j(x_i)$. The general expression (7.5) for the χ^2 can then be written in matrix notation,

$$\begin{aligned} \chi^2 &= (\mathbf{y} - \boldsymbol{\lambda})^T V^{-1} (\mathbf{y} - \boldsymbol{\lambda}) \\ &= (\mathbf{y} - A\boldsymbol{\theta})^T V^{-1} (\mathbf{y} - A\boldsymbol{\theta}), \end{aligned} \tag{7.8}$$

where $\mathbf{y} = (y_1, \ldots, y_N)$ is the vector of measured values, and $\boldsymbol{\lambda} = (\lambda_1, \ldots, \lambda_N)$ contains the predicted values $\lambda_i = \lambda(x_i; \boldsymbol{\theta})$. In matrix equations, $\mathbf{y}$ and $\boldsymbol{\lambda}$ are understood to be column vectors, and the superscript T indicates a transposed (i.e. row) vector.

To find the minimum χ^2 we set its derivatives with respect to the parameters θ_i equal to zero,

$$\nabla \chi^2 = -2(A^T V^{-1} \mathbf{y} - A^T V^{-1} A\boldsymbol{\theta}) = 0. \tag{7.9}$$

Providing the matrix $A^T V^{-1} A$ is not singular, this can be solved for the estimators $\hat{\boldsymbol{\theta}}$,

$$\hat{\boldsymbol{\theta}} = (A^T V^{-1} A)^{-1} A^T V^{-1} \mathbf{y} \equiv B\, \mathbf{y}. \tag{7.10}$$

That is, the solutions $\hat{\boldsymbol{\theta}}$ are linear functions of the original measurements $\mathbf{y}$. Using error propagation to find the covariance matrix for the estimators $U_{ij} = \text{cov}[\hat{\theta}_i, \hat{\theta}_j]$ gives

$$U = B V B^T = (A^T V^{-1} A)^{-1}. \tag{7.11}$$

Equivalently, the inverse covariance matrix U^{-1} can be expressed as

$$(U^{-1})_{ij} = \frac{1}{2} \left[\frac{\partial^2 \chi^2}{\partial \theta_i \partial \theta_j} \right]_{\boldsymbol{\theta} = \hat{\boldsymbol{\theta}}}. \tag{7.12}$$

Note that equation (7.12) coincides with the RCF bound for the inverse covariance matrix when the y_i are Gaussian distributed, where one has $\log L = -\chi^2/2$, cf. Sections 6.6, 7.1.

For the case of $\lambda(x; \boldsymbol{\theta})$ linear in the parameters $\boldsymbol{\theta}$, the χ^2 is quadratic in $\boldsymbol{\theta}$:

$$\chi^2(\boldsymbol{\theta}) = \chi^2(\hat{\boldsymbol{\theta}}) + \frac{1}{2} \sum_{i,j=1}^{m} \left[\frac{\partial^2 \chi^2}{\partial \theta_i \partial \theta_j} \right]_{\boldsymbol{\theta} = \hat{\boldsymbol{\theta}}} (\theta_i - \hat{\theta}_i)(\theta_j - \hat{\theta}_j). \tag{7.13}$$

Combining this with the expression for the variance given by equation (7.12) yields the contours in parameter space whose tangents are at $\hat{\theta}_i \pm \hat{\sigma}_i$, corresponding to a one standard deviation departure from the LS estimates:

$$\chi^2(\boldsymbol{\theta}) = \chi^2(\hat{\boldsymbol{\theta}}) + 1 = \chi^2_{\min} + 1. \tag{7.14}$$

This corresponds directly to the contour of constant likelihood seen in connection with the maximum likelihood problem of Section 6.11. If the function $\lambda(x; \boldsymbol{\theta})$ is not linear in the parameters, then the contour defined by equation (7.14) is not in general elliptical, and one can no longer obtain the standard deviations from the tangents. It defines a region in parameter space, however, which can be interpreted as a **confidence region**, the size of which reflects the statistical uncertainty of the fitted parameters. The concept of confidence regions will be defined more precisely in Chapter 9. One should note, however, that the confidence level of the region defined by (7.14) depends on the number of parameters fitted: 6.83% for one parameter, 39.4% for two, 19.9% for three, etc. (cf. Section 9.7).

7.3   Least squares fit of a polynomial

As an example of the least squares method consider the data shown in Fig. 7.2, consisting of five values of a quantity y measured with errors Δy at different values of x. Assume the measured values y_i each come from a Gaussian distribution centered around λ_i (which is unknown) with a standard deviation $\sigma_i = \Delta y_i$ (assumed known). As a hypothesis for $\lambda(x; \boldsymbol{\theta})$ one might try a polynomial of order m (i.e. $m + 1$ parameters),

$$\lambda(x;\theta_0,\ldots,\theta_m) = \sum_{j=0}^{m} x^j\,\theta_j. \qquad (7.15)$$

This is a special case of the linear least-squares fit described in Section 7.2 with the coefficient functions $a_j(x)$ equal to powers of x. Figure 7.2 shows the LS fit result for polynomials of order 0, 1 and 4. The zero-order polynomial is simply the average of the measured values, with each point weighted inversely by the square of its error. This hypothesis gives $\hat{\theta}_0 = 2.66 \pm 0.13$ and $\chi^2 = 45.5$ for four degrees of freedom (five points minus one free parameter). The data are better described by a straight-line fit (first-order polynomial) giving $\hat{\theta}_0 = 0.93 \pm 0.30$, $\hat{\theta}_1 = 0.68 \pm 0.10$ and $\chi^2 = 3.99$ for three degrees of freedom. Since there are only five data points, the fourth-order polynomial (with five free parameters) goes exactly through every point yielding a χ^2 of zero. The use of the χ^2 value to evaluate the goodness-of-fit will be discussed in Section 7.5.

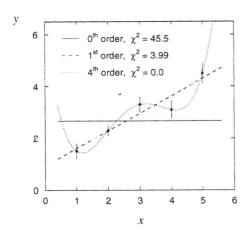

Fig. 7.2 Least squares fits of polynomials of order 0, 1 and 4 to five measured values.

As in the case of the maximum likelihood method, the statistical errors and covariances of the estimators can be estimated in several ways. All are related to the *change* in the χ^2 as the parameters are moved away from the values for which χ^2 is a minimum. Figure 7.3(a) shows the χ^2 as a function of θ_0 for the case of the zero-order polynomial. The χ^2 curve is a parabola, since the hypothesized fit function is linear in the parameter θ_0 (see equation (7.13)). The variance of the LS estimator $\hat{\theta}_0$ can be evaluated by any of the methods discussed in Section 7.2: from the change in the parameter necessary to increase the minimum χ^2 by one, from the curvature (second derivative) of the parabola at its minimum, or directly from equation (7.11).

Figure 7.3(b) shows a contour of $\chi^2 = \chi^2_{\min} + 1$ for the first-order polynomial (two-parameter) fit. From the inclination and width of the ellipse one can see that the estimators $\hat{\theta}_0$ and $\hat{\theta}_1$ are negatively correlated. Equation 7.11 gives

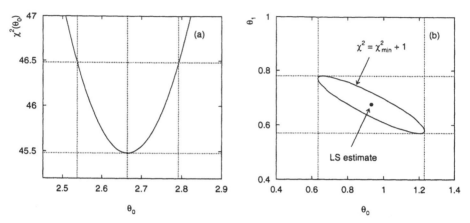

Fig. 7.3 (a) The χ^2 as a function of θ_0 for the zero-order polynomial fit shown in Fig. 7.2. The horizontal lines indicate $\chi^2_{\min}$ and $\chi^2_{\min}+1$. The corresponding θ_0 values (vertical lines) are the LS estimate $\hat{\theta}_0$ and $\hat{\theta}_0 \pm \hat{\sigma}_{\hat{\theta}_0}$. (b) The LS estimates $\hat{\theta}_0$ and $\hat{\theta}_1$ for the first-order polynomial fit in Fig. 7.2. The tangents to the contour $\chi^2(\hat{\theta}_0, \hat{\theta}_1) = \chi^2_{\min} + 1$ correspond to $\hat{\theta}_0 \pm \hat{\sigma}_{\hat{\theta}_0}$ and $\hat{\theta}_1 \pm \hat{\sigma}_{\hat{\theta}_1}$.

$$\hat{\sigma}_{\hat{\theta}_0} = \sqrt{\hat{U}_{00}} = 0.30$$

$$\hat{\sigma}_{\hat{\theta}_1} = \sqrt{\hat{U}_{11}} = 0.10$$

$$\widehat{\text{cov}}[\hat{\theta}_0, \hat{\theta}_1] = \hat{U}_{01} = -0.028,$$

corresponding to a correlation coefficient of $r = -0.90$. As in the case of maximum likelihood, the standard deviations correspond to the tangents of the ellipse, and the correlation coefficient to its width and angle of inclination (see equations (6.31) and (6.32)).

Since the two estimators $\hat{\theta}_0$ and $\hat{\theta}_1$ have a strong negative correlation, it is important to include the covariance, or equivalently the correlation coefficient, when reporting the results of the fit. Recall from Section 1.7 that one can always define two new quantities, $\hat{\eta}_0$ and $\hat{\eta}_1$, from the original $\hat{\theta}_0$ and $\hat{\theta}_1$ by means of an orthogonal transformation such that $\text{cov}[\hat{\eta}_0, \hat{\eta}_1] = 0$. However, although it is generally easier to deal with uncorrelated quantities, the transformed parameters may not have as direct an interpretation as the original ones.

7.4   Least squares with binned data

In the previous examples, the function relating the 'true' values λ to the variable x was not necessarily a p.d.f. for x, but an arbitrary function. It can, however, be a p.d.f., or it can be proportional to one. Suppose, for example, one has n

observations of a random variable x from which one makes a histogram with
N bins. Let y_i be the number of entries in bin i and $f(x;\boldsymbol{\theta})$ be a hypothesized
p.d.f. for which one would like to estimate the parameters $\boldsymbol{\theta} = (\theta_1,\ldots,\theta_m)$. The
number of entries predicted in bin i, $\lambda_i = E[y_i]$, is then

$$\lambda_i(\boldsymbol{\theta}) = n \int_{x_i^{\min}}^{x_i^{\max}} f(x;\boldsymbol{\theta})dx = np_i(\boldsymbol{\theta}), \tag{7.16}$$

where $x_i^{\min}$ and $x_i^{\max}$ are the bin limits and $p_i(\boldsymbol{\theta})$ is the probability to have an
entry in bin i. The parameters $\boldsymbol{\theta}$ are found by minimizing the quantity

$$\chi^2(\boldsymbol{\theta}) = \sum_{i=1}^{N} \frac{(y_i - \lambda_i(\boldsymbol{\theta}))^2}{\sigma_i^2}, \tag{7.17}$$

where σ_i^2 is the variance of the number of entries in bin i. Note that here the
function $f(x;\boldsymbol{\theta})$ is normalized to one, since it is a p.d.f., and the function that
is fitted to the histogram is $\lambda_i(\boldsymbol{\theta})$.

If the mean number of entries in each bin is small compared to the total
number of entries, the contents of each bin are approximately Poisson distributed.
The variance is therefore equal to the mean (see equation (2.11)) so that equation
(7.17) becomes

$$\chi^2(\boldsymbol{\theta}) = \sum_{i=1}^{N} \frac{(y_i - \lambda_i(\boldsymbol{\theta}))^2}{\lambda_i(\boldsymbol{\theta})} = \sum_{i=1}^{N} \frac{(y_i - np_i(\boldsymbol{\theta}))^2}{np_i(\boldsymbol{\theta})}. \tag{7.18}$$

An alternative method is to approximate the variance of the number of entries
in bin i by the number of entries actually observed y_i, rather than by the predicted number $\lambda_i(\boldsymbol{\theta})$. This is the so-called **modified least-squares method** (MLS
method) for which one minimizes

$$\chi^2(\boldsymbol{\theta}) = \sum_{i=1}^{N} \frac{(y_i - \lambda_i(\boldsymbol{\theta}))^2}{y_i} = \sum_{i=1}^{N} \frac{(y_i - np_i(\boldsymbol{\theta}))^2}{y_i}. \tag{7.19}$$

This may be easier to deal with computationally, but has the disadvantage that
the errors may be poorly estimated (or χ^2 may even be undefined) if any of the
bins contain few or no entries.

When using the LS method for fitting to a histogram one should be aware
of the following potential problem. Often instead of using the observed total
number of entries n to obtain λ_i from equation (7.16), an additional adjustable
parameter ν is introduced as a normalization factor. The predicted number of
entries in bin i, $\lambda_i(\boldsymbol{\theta},\nu) = E[y_i]$, then becomes

$$\lambda_i(\boldsymbol{\theta},\nu) = \nu \int_{x_i^{\min}}^{x_i^{\max}} f(x;\boldsymbol{\theta})dx = \nu p_i(\boldsymbol{\theta}). \tag{7.20}$$

This step would presumably be taken in order to eliminate the need to count
the number of entries n. One can easily show, however, that introducing an

adjustable normalization parameter leads to an incorrect estimate of the total number of entries. Consider the LS case where the variances are taken from the predicted number of entries ($\sigma_i^2 = \lambda_i$). Using equation (7.20) for λ_i and setting the derivative of χ^2 with respect to ν equal to zero gives the estimator

$$\hat{\nu}_{\mathrm{LS}} = n + \frac{\chi^2}{2}. \tag{7.21}$$

For the MLS case ($\sigma_i^2 = y_i$) one obtains

$$\hat{\nu}_{\mathrm{MLS}} = n - \chi^2. \tag{7.22}$$

Since one expects a contribution to χ^2 on the order of one per bin, the relative error in the number of entries is typically $N/2n$ too high (LS) or N/n too low (MLS). If one takes as a rule of thumb that each bin should have at least five entries, one could have an (unnecessary) error in the normalization of 10–20%.

   Although the bias introduced may be smaller than the corresponding statistical error, a result based on the average of such fits could easily be wrong by an amount larger than the statistical error of the average. Therefore, one should determine the normalization directly from the number of entries. If this is not practical (e.g. because of software constraints) one should at least be aware that a potential problem exists, and the bin size should be chosen such that the bias introduced is acceptably small.

   The least squares method with binned data can be compared to the maximum likelihood technique of Section 6.10. There the joint p.d.f. for the bin contents y_i was taken to be a multinomial distribution, or alternatively each y_i was regarded as a Poisson random variable. Recall that in the latter case, where the expected total number of entries ν was treated as an adjustable parameter, the correct value $\hat{\nu} = n$ was automatically found (cf. Sections 6.9, 6.10). Furthermore it has been pointed out in [Ead71] (Section 8.4.5 and references therein) that the variances of ML estimators converge faster to the minimum variance bound than LS or MLS estimators, giving an additional reason to prefer the ML method for histogram fitting.

   As an example consider the histograms shown in Fig. 7.4, which contain 400 entries in 20 bins. The data were generated by the Monte Carlo method in the interval $[0, 2]$. The p.d.f. used was a linear combination of a Gaussian and an exponential given by

$$f(x; \theta, \mu, \sigma, \xi) = \theta \, \frac{\exp\left(\frac{-(x-\mu)^2}{2\sigma^2}\right)}{\int_0^2 \exp\left(\frac{-(x'-\mu)^2}{2\sigma^2}\right) dx'} + (1 - \theta) \frac{e^{-x/\xi}}{\xi(1 - e^{-2/\xi})}, \tag{7.23}$$

with $\theta = 0.5$, $\mu = 1$, $\xi = 4$ and $\sigma = 0.35$. Assume that μ, σ and ξ are known, and that one would like to determine the number of entries contributing to the Gaussian and exponential components, $\nu_{\mathrm{Gauss}} = \nu\theta$ and $n_{\mathrm{exp}} = \nu(1 - \theta)$.

Figure 7.4(a) shows the fit results where ν was treated as a free parameter. For the case $\sigma_i^2 = \lambda_i$ (LS) one obtains $\chi^2 = 17.1$ and $\hat{\nu}_{LS} = 408.5$, as expected from equation (7.21), and $\hat{\theta} = 0.498 \pm 0.056$. For the case $\sigma_i^2 = y_i$ (MLS) one obtains $\chi^2 = 17.8$, $\hat{\nu}_{MLS} = 382.2 \pm 19.5$, which is in accordance with equation (7.22), and $\hat{\theta} = 0.551 \pm 0.062$. One clearly sees from the figure that the areas under the two fitted curves are different, and this leads to different estimated numbers of entries corresponding to the Gaussian and exponential components.

Figure 7.4(b) shows fit results where ν is treated more correctly. For the curve labeled LS, the variances have been taken to be $\sigma_i^2 = \lambda_i$ and the total number of entries ν has been fixed to the true number of entries, $\nu = n = 400$. This results in $\chi^2 = 17.3$ and $\hat{\theta} = 0.496 \pm 0.055$. Also shown in Fig. 7.4(b) is the result of an ML fit with $\hat{\theta} = 0.514 \pm 0.057$, where the likelihood function is given by the product of Poisson distributions. As shown in Section 6.10, the ML estimator automatically gives the correct number of entries. The goodness-of-fit can be evaluated in the ML case using the statistic (6.52). This gives $\chi_P^2 = 17.6$, similar to the χ^2 from the LS fit. Although the standard deviations of $\hat{\theta}$ are similar in all of the techniques shown, the fits shown in Fig. 7.4(b) are to be preferred, since there the total number of entries is correct.

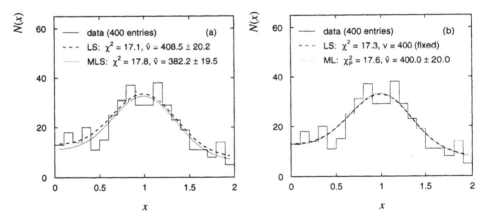

Fig. 7.4 (a) Fits to Monte Carlo data generated according to equation (7.23) where the total number of entries ν is treated as an adjustable parameter. (b) Fit results using the LS method with the total number of entries fixed to the true number and using the method of maximum likelihood (see text).

7.5   Testing goodness-of-fit with χ^2

If the measured values y_i are Gaussian, the resulting estimators coincide with the ML estimators, as seen in Section 7.1. Furthermore, the χ^2 value can be used as a test of how likely it is that the hypothesis, if true, would yield the observed data.

The quantity $(y_i - \lambda(x_i; \boldsymbol{\theta}))/\sigma_i$ is a measure of the deviation between the ith measurement y_i and the function $\lambda(x_i; \boldsymbol{\theta})$, so χ^2 is a measure of total agreement between observed data and hypothesis. It can be shown that if

(1) the $y_i, i = 1, \ldots, N$, are independent Gaussian random variables with known variances, σ_i^2 (or are distributed according to an N-dimensional Gaussian with known covariance matrix V);

(2) the hypothesis $\lambda(x; \theta_1, \ldots, \theta_m)$ is linear in the parameters θ_i; and

(3) the functional form of the hypothesis is correct,

then the minimum value of χ^2 defined by equation (7.3) (or for correlated y_i by equation (7.5)) is distributed according to the χ^2 distribution with $N - m$ degrees of freedom as defined in Section 2.7, equation (2.34). We have already encountered a special case of this in Section 4.7, where no parameters were determined from the data.

As seen in Section 2.7, the expectation value of a random variable z from the χ^2 distribution is equal to the number of degrees of freedom. One often quotes therefore the χ^2 divided by the number of degrees of freedom n_d (the number of data points minus the number of independent parameters) as a measure of goodness-of-fit. If it is near one, then all is as expected. If it is much less than one, then the fit is better than expected given the size of the measurement errors. This is not bad in the sense of providing evidence against the hypothesis, but it is usually grounds to check that the errors σ_i have not been overestimated or are not correlated.

If χ^2/n_d is much larger than one, then there is some reason to doubt the hypothesis. As discussed in Section 4.5, one often quotes a significance level (P-value) for a given χ^2, which is the probability that the hypothesis would lead to a χ^2 value worse (i.e. greater) than the one actually obtained. That is,

$$P = \int_{\chi^2}^{\infty} f(z; n_d)dz, \tag{7.24}$$

where $f(z; n_d)$ is the χ^2 distribution for n_d degrees of freedom. Values can be computed numerically (e.g. with the routine **PROB** in [CER97]) or looked up in standard graphs or tables (e.g. references [PDG96, Bra92]). The P-value at which one decides to reject a hypothesis is subjective, but note that underestimated errors, σ_i, can cause a correct hypothesis to give a bad χ^2.

For the polynomial fit considered in Section 7.3, one obtained for the straight-line fit $\chi^2 = 3.99$ for three degrees of freedom (five data points minus two free parameters). Computing the significance level using equation (7.24) gives $P = 0.263$. That is, if the true function $\lambda(x)$ were a straight line and if the experiment were repeated many times, each time yielding values for $\hat{\theta}_0$, $\hat{\theta}_1$ and χ^2, then one would expect the χ^2 values to be worse (i.e. higher) than the one actually obtained ($\chi^2 = 3.99$) in 26.3% of the cases. This can be checked by performing a large number of Monte Carlo experiments where the 'true' parameters θ_0 and θ_1 are taken from the results of the real experiment, and a 'measured' value for each data point is generated from a Gaussian of width σ given by the corresponding errors. Figure 7.5 shows a normalized histogram of the χ^2 values from 1000 simulated experiments along with the predicted χ^2 distribution for three degrees of freedom.

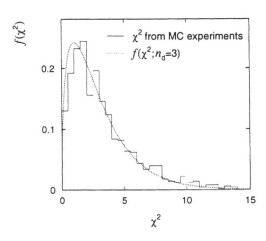

Fig. 7.5 Normalized histogram of χ^2 values from 1000 Monte Carlo experiments along with the predicted χ^2 distribution for three degrees of freedom.

The fit to the horizontal line gave $\chi^2 = 45.5$ for four degrees of freedom. The corresponding significance level is $P = 3.1 \times 10^{-9}$. If the horizontal-line hypothesis were true, one would expect a χ^2 as high or higher than the one obtained in only three out of a billion experiments, so this hypothesis can safely be ruled out. In computing the P-value it was assumed that the standard deviations σ_i (or for correlated measurements the covariance matrix V) were known. One should keep in mind that underestimated measurement errors σ_i or incorrect treatment of correlations can cause a correct hypothesis to result in a large χ^2.

One should keep in mind the distinction between having small statistical errors and having a good (i.e. small) χ^2. The statistical errors are related to the *change* in χ^2 when the parameters are varied away from their fitted values, and not to the absolute value of χ^2 itself. From equation (7.11) one can see that the covariance matrix of the estimators U depends only on the coefficient functions $a_j(x)$ (i.e. on the composite hypothesis $\lambda(x; \boldsymbol{\theta})$) and on the covariance matrix V of the original measurements, but is independent of the measured values y_i.

The standard deviation $\sigma_{\hat{\theta}}$ of an estimator $\hat{\theta}$ is a measure of how widely estimates would be distributed if the experiment were to be repeated many times. If the functional form of the hypothesis is incorrect, however, then the estimate $\hat{\theta}$ can still differ significantly from the true value θ, which would be defined in the true composite hypothesis. That is, if the form of the hypothesis is incorrect, then a small standard deviation (statistical error) is not sufficient to imply a small uncertainty in the estimate of the parameter.

To demonstrate this point, consider the fit to the horizontal line done in Section 7.3, which yielded the estimate $\hat{\theta}_0 = 2.66 \pm 0.13$ and $\chi^2 = 45.5$ for four degrees of freedom. Figure 7.6 shows a set of five data points with the same x values and the same errors, Δy, but with different y values. A fit to a horizontal line gives $\hat{\theta}_0 = 2.84 \pm 0.13$ and $\chi^2 = 4.48$. The error on $\hat{\theta}_0$ stays the same, but the χ^2 value is now such that the horizontal-line hypothesis provides a good description of the data. The χ^2 vs. θ_0 curves for the two cases have the same curvature, but one is simply shifted vertically with respect to the other by a constant offset.

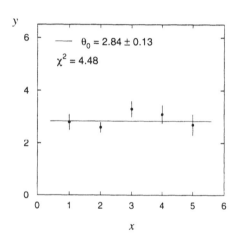

Fig. 7.6 Least squares fit of a polynomial of order 0 to data with the same x values and errors as shown in Fig. 7.2, but with different y values. Although the χ^2 value is much smaller than in the previous example, the error of $\hat{\theta}_0$ remains the same.

7.6   Combining measurements with least squares

A special case of the LS method is often used to combine a number of measurements of the same quantity. Suppose that a quantity of unknown true value λ has been measured N times (e.g. in N different experiments) yielding independent values y_i and estimated errors (standard deviations) σ_i for $i = 1, \dots, N$. Since one assumes that the true value is the same for all the measurements, the value λ is a constant, i.e. the function $\lambda(x)$ is a constant, and thus the variable x does not actually appear in the problem. Equation (7.3) becomes

$$\chi^2(\lambda) = \sum_{i=1}^{N} \frac{(y_i - \lambda)^2}{\sigma_i^2}, \tag{7.25}$$

where λ plays the role of the parameter θ. Setting the derivative of χ^2 with respect to λ equal to zero and solving for λ gives the LS estimator $\hat{\lambda}$,

$$\hat{\lambda} = \frac{\sum_{i=1}^{N} y_i/\sigma_i^2}{\sum_{j=1}^{N} 1/\sigma_j^2}, \tag{7.26}$$

which is the well-known formula for a weighted average. From the second derivative of χ^2 one obtains the variance of $\hat{\lambda}$ (see equation (7.12)),

$$V[\hat{\lambda}] = \frac{1}{\sum_{i=1}^{N} 1/\sigma_i^2}. \tag{7.27}$$

From equation (7.27) one sees that the variance of the weighted average is smaller than any of the variances of the individual measurements. Furthermore, if one of the measured y_i has a much smaller variance than the rest, then this measurement will dominate both in the value and variance of the weighted average.

This procedure can easily be generalized to the case where the measurements y_i are not independent. This would occur, for example, if they are based at least in part on the same data. Assuming that the covariance matrix V for the N measurements is known, equation (7.25) can be rewritten using the more general definition of the χ^2 (equation (7.5)),

$$\chi^2(\lambda) = \sum_{i,j=1}^{N} (y_i - \lambda)(V^{-1})_{ij}(y_j - \lambda). \tag{7.28}$$

The LS estimator for λ is found as usual by setting the derivative of $\chi^2(\lambda)$ with respect to λ equal to zero. As in the case of uncorrelated measurements, the resulting estimator is a linear combination of the measured y_i,

$$\hat{\lambda} = \sum_{i=1}^{N} w_i y_i, \tag{7.29}$$

with the weights w_i now given by

$$w_i = \frac{\sum_{j=1}^{N} (V^{-1})_{ij}}{\sum_{k,l=1}^{N} (V^{-1})_{kl}}. \tag{7.30}$$

This reduces of course to equation (7.26) for the case of uncorrelated measurements, where $(V^{-1})_{ij} = \delta_{ij}/\sigma_i^2$.

Note that the weights w_i sum to unity,

$$\sum_{i=1}^{N} w_i = \frac{\sum_{i,j=1}^{N} (V^{-1})_{ij}}{\sum_{k,l=1}^{N} (V^{-1})_{kl}} = 1. \tag{7.31}$$

Assuming that the individual measurements y_i are all unbiased estimates of λ, this implies that the estimator $\hat{\lambda}$ is also unbiased,

$$E[\hat{\lambda}] = \sum_{i=1}^{N} w_i E[y_i] = \lambda \sum_{i=1}^{N} w_i = \lambda. \tag{7.32}$$

This is true for any choice of the weights as long as one has $\sum_{i=1}^{N} w_i = 1$. One can show that the particular weights given by the LS prescription (7.30) lead to the unbiased estimator $\hat{\lambda}$ with the smallest possible variance. On the one hand, this follows from the Gauss–Markov theorem, which holds for all LS estimators. Equivalently, one could simply assume the form of a weighted average (7.29), require $\sum_{i=1}^{N} w_i = 1$, and determine the weights such that the variance of $\hat{\lambda}$ is a minimum. By error propagation, equation (1.53), one obtains the variance

$$V[\hat{\lambda}] = \sum_{i,j=1}^{N} w_i V_{ij} w_j, \tag{7.33}$$

or in matrix notation, $V[\hat{\lambda}] = \mathbf{w}^T V \mathbf{w}$, where $\mathbf{w}$ is a column vector containing the N weights and $\mathbf{w}^T$ is the corresponding transposed (i.e. row) vector. By replacing w_N by $1 - \sum_{i=1}^{N-1} w_i$, and setting the derivatives of equation (7.33) with respect to the first $N-1$ of the w_i equal to zero, one obtains exactly the weights given by equation (7.30). (Alternatively, a Lagrange multiplier can be used to impose the constraint $\sum_{i=1}^{N} w_i = 1$.)

As an example, consider two measurements y_1 and y_2, with the covariance matrix

$$V = \begin{pmatrix} \sigma_1^2 & \rho\sigma_1\sigma_2 \\ \rho\sigma_1\sigma_2 & \sigma_2^2 \end{pmatrix}, \tag{7.34}$$

where $\rho = V_{12}/(\sigma_1\sigma_2)$ is the correlation coefficient. The inverse covariance matrix is then given by

$$V^{-1} = \frac{1}{1-\rho^2} \begin{pmatrix} \frac{1}{\sigma_1^2} & \frac{-\rho}{\sigma_1\sigma_2} \\ \frac{-\rho}{\sigma_1\sigma_2} & \frac{1}{\sigma_2^2} \end{pmatrix} \tag{7.35}$$

Using this in equations (7.29) and (7.30) yields the weighted average,

$$\hat{\lambda} = wy_1 + (1-w)y_2, \tag{7.36}$$

with

$$w = \frac{\sigma_2^2 - \rho\sigma_1\sigma_2}{\sigma_1^2 + \sigma_2^2 - 2\rho\sigma_1\sigma_2}. \tag{7.37}$$

From equation (7.33) the variance of $\hat{\lambda}$ is found to be

$$V[\hat{\lambda}] = \frac{(1-\rho^2)\sigma_1^2\sigma_2^2}{\sigma_1^2 + \sigma_2^2 - 2\rho\sigma_1\sigma_2} = \sigma^2, \tag{7.38}$$

or equivalently one has

$$\frac{1}{\sigma^2} = \frac{1}{1 - \rho^2} \left[\frac{1}{\sigma_1^2} + \frac{1}{\sigma_2^2} - \frac{2\rho}{\sigma_1 \sigma_2} \right]. \tag{7.39}$$

The presence of correlations has some interesting consequences; see e.g. [Lyo88, Dag94]. From equation (7.39) one has that the change in the inverse variance due to the second measurement y_2 is

$$\frac{1}{\sigma^2} - \frac{1}{\sigma_1^2} = \frac{1}{1 - \rho^2} \left(\frac{\rho}{\sigma_1} - \frac{1}{\sigma_2} \right)^2. \tag{7.40}$$

This is always greater than or equal to zero, which is to say that the second measurement always helps to decrease σ^2, or at least it never hurts. The change in the variance is zero when $\rho = \sigma_1/\sigma_2$. This includes the trivial case with $\rho = 1$ and $\sigma_1 = \sigma_2$, i.e. the same measurement is considered twice.

If $\rho > \sigma_1/\sigma_2$, the weight w becomes negative, which means that the weighted average does not lie between y_1 and y_2. This comes about because for a large positive correlation between y_1 and y_2, both values are likely to lie on the same side of the true value λ. This is a sufficiently surprising result that it is worth examining more closely in the following example.

7.6.1 An example of averaging correlated measurements

Consider measuring the length of an object with two rulers made of different substances, so that the thermal expansion coefficients are different. Suppose both rulers have been calibrated to give accurate results at a temperature T_0, but at any other temperature, a corrected estimate y of the true (unknown) length λ must be obtained using

$$y_i = L_i + c_i(T - T_0). \tag{7.41}$$

Here the index i refers to ruler 1 or 2, L_i is the uncorrected measurement, c_i is the expansion coefficient, and T is the temperature, which must be measured. We will treat the measured temperature as a random variable with standard deviation σ_T, and we assume that T is the same for the two measurements, i.e. they are carried out together. The uncorrected measurements L_i are treated as random variables with standard deviations σ_{L_i}. Assume that the c_i, σ_{L_i} and σ_T are known.

In order to obtain the weighted average of y_1 and y_2, we need their covariance matrix. The variances of the corrected measurements $V[y_i] = \sigma_i^2$ are

$$\sigma_i^2 = \sigma_{L_i}^2 + c_i^2 \sigma_T^2. \tag{7.42}$$

Assume that the measurements are unbiased, i.e. $E[y_i] = \lambda$ and $E[T] = T_{\text{true}}$, where T_{true} is the true (unknown) temperature. One then has the expectation values $E[L_i] = \lambda - c_i(T_{\text{true}} - T_0)$, $E[T^2] = \sigma_T^2 + T_{\text{true}}^2$, and $E[L_i, L_j] = \delta_{ij} \sigma_{L_i}^2$. From these the covariance $V_{12} = \text{cov}[y_1, y_2]$ is found to be

$$
\begin{aligned}
V_{12} &= E[y_1 y_2] - \lambda^2 \\
&= E[(L_1 + c_1(T - T_0))(L_2 + c_2(T - T_0))] - \lambda^2 \\
&= c_1 c_2 \sigma_T^2,
\end{aligned}
\tag{7.43}
$$

and the correlation coefficient is

$$
\rho = \frac{V_{12}}{\sigma_1 \sigma_2} = \frac{c_1 c_2 \sigma_T^2}{\sqrt{(\sigma_{L_1}^2 + c_1^2 \sigma_T^2)(\sigma_{L_2}^2 + c_2^2 \sigma_T^2)}}.
\tag{7.44}
$$

The weighted average (7.26) is thus

$$
\hat{\lambda} = \frac{(\sigma_{L_2}^2 + (c_2^2 - c_1 c_2)\sigma_T^2)\, y_1 + (\sigma_{L_1}^2 + (c_1^2 - c_1 c_2)\sigma_T^2)\, y_2}{\sigma_{L_1}^2 + \sigma_{L_2}^2 + (c_1 - c_2)^2 \sigma_T^2}.
\tag{7.45}
$$

If the error in the temperature σ_T is negligibly small, then ρ goes to zero, and $\hat{\lambda}$ will lie between y_1 and y_2. If, however, the standard deviations σ_{L_i} are very small and the error in the temperature σ_T is large, then the correlation coefficient (7.44) approaches unity, and the weight (7.37) becomes negative. For the extreme case of $\rho = 1$, (7.45) becomes

$$
\hat{\lambda} = \frac{-c_1}{c_1 - c_2} y_1 + \frac{c_2}{c_1 - c_2} y_2,
\tag{7.46}
$$

which has a variance of zero, cf. equation (7.38).

The reason for this outcome is illustrated in Fig. 7.7. The two diagonal bands represent the possible values of the corrected length y_i, given measured lengths L_i, as a function of the temperature. If the L_i are known very accurately, and yet if y_1 and y_2 differ by quite a bit, then the only available explanation is that the true temperature must be different from the measured value T. The weighted average $\hat{\lambda}$ is thus pulled towards the point where the bands of $y_1(T)$ and $y_2(T)$ cross.

The problem is equivalent to performing a least squares fit for the length λ and the true temperature T_{true}, given the uncorrelated measurements L_1, L_2 and T. The LS estimator for the temperature is found to be

$$
\hat{T} = T - \frac{(c_1 - c_2)\,(y_1 - y_2)\,\sigma_T^2}{\sigma_{L_1}^2 + \sigma_{L_2}^2 + (c_1 - c_2)^2 \sigma_T^2}.
\tag{7.47}
$$

The knowledge that the temperature was the same for both measurements and the assumption that the error in T was at least in part responsible for the discrepancy between y_1 and y_2 are exploited to obtain estimates of both λ and T_{true}.

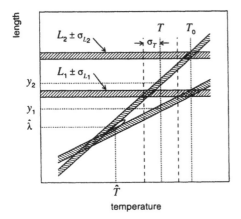

Fig. 7.7 The horizontal bands indicate the measured lengths L_i of an object with two different rulers, and the diagonal bands show the corrected lengths as a function of the temperature. The measured temperature T leads to the corrected lengths y_1 and y_2. The error ellipse corresponds to $\chi^2 = \chi^2_{min} + 1$ and is centered about the estimates for the length $\hat{\lambda}$ and the temperature $\hat{T}$.

One should ask to what extent such a procedure can be trusted in practice. The correlation stems from the fact that the individual measurements are based in part on a common measurement, here the temperature, which itself is a random quantity. Averages of highly correlated quantities should be treated with caution, since a small error in the covariance matrix for the individual measurements can lead to a large error in the average $\hat{\lambda}$, as well as an incorrect estimation of its variance. This is particularly true if the magnitudes of correlated uncertainties are overestimated.

In the example above, if the correction to the temperature $\Delta T = \hat{T} - T$ turns out to be large compared to σ_T, then this means that our assumptions about the measurements are probably incorrect. This would be reflected in a large value of the minimized χ^2, and a correspondingly small P-value. This is the case in Fig. 7.7, which gives $\chi^2 = 10.7$ for one degree of freedom. We would then have to revise the model, perhaps reevaluating $\sigma^2_{L_1}$, $\sigma^2_{L_2}$ and σ^2_T, or the relationship between the measured and corrected lengths. The correct conclusion in the example here would be that because of the large χ^2, the two measurements y_1 and y_2 are incompatible, and the average is therefore not reliable.

One could imagine, for example, that the reason for the large χ^2 is that the standard deviations of the L_i are underestimated. If they would be taken larger by a factor of four, then one would obtain $\chi^2 = 1.78$ and the $\hat{\lambda}$ would lie between y_1 and y_2. Suppose instead, however, that one were to take σ_T larger by a factor of four. This would cause the χ^2 to drop to 0.97, so that this would not provide any evidence against the modeling of the system. The error ellipse for $\hat{\lambda}$ and $\hat{T}$ (i.e. the contour of $\chi^2 = \chi^2_{min} + 1$) would shift closer to the point where the two bands cross, but would remain approximately the same size. If the larger value of σ_T is known to be correct, and the modeling of the measuring devices is reliable, then the result is perfectly acceptable. If, however, σ_T is simply overestimated in an attempt to be conservative, then one would be led to an erroneous result.

7.6.2   Determining the covariance matrix

In order to apply the LS technique to average results, it is necessary to know the covariance matrix V. In simple cases one may be able to determine the covariance directly, e.g. if two measurements simply use some, but not all, of the same data. Consider, for example, a number of independent observations of a quantity x ('events'), where x itself has a variance σ^2. Suppose that the first measurement y_1 is constructed as the sample mean of n independent events,

$$y_1 = \frac{1}{n} \sum_{i=1}^{n} x_i, \qquad (7.48)$$

and a second measurement y_2 uses m values of x,

$$y_2 = \frac{1}{m} \sum_{i=1}^{m} x_i. \qquad (7.49)$$

Suppose further that c of the events are common to the two sets, i.e. they represent the same observations. Using $\mathrm{cov}[x_i, x_j] = \delta_{ij}\sigma^2$, the covariance $\mathrm{cov}[y_1, y_2]$ is found to be

$$\mathrm{cov}[y_1, y_2] = \frac{c\sigma^2}{nm}, \qquad (7.50)$$

or equivalently $\rho = c/\sqrt{nm}$. That is, if $c = 0$, the two data sets do not overlap and one has $\rho = 0$; if $c \approx n \approx m$, then the measurements use almost all of the same data, and hence are highly correlated.

In this example the variances of the individual measurements are $\sigma_1 = \sigma/\sqrt{n}$ and $\sigma_2 = \sigma/\sqrt{m}$. One has thus $\sigma_1/\sigma_2 = \sqrt{m/n}$ and therefore

$$\rho = \frac{c}{\sqrt{nm}} = \frac{\sigma_1}{\sigma_2} \frac{c}{m} \leq \frac{\sigma_1}{\sigma_2}. \qquad (7.51)$$

In this case, therefore, a negative weight cannot occur and the average always lies between y_1 and y_2.

Often it is more difficult to determine the correlation between two estimators. Even if two data samples overlap completely, different estimators, $y_1, \ldots, y_N$, may depend on the data in different ways. Determining the covariance matrix can therefore be difficult, and may require simulating a large number of experiments with the Monte Carlo method, determining $y_1, \ldots, y_N$ for each experiment and then estimating the covariance of each pair of measurements with equation (5.11). An example with mean particle lifetimes treated in this way is given in reference [Lyo88].

In situations where even one simulated experiment might involve a long Monte Carlo calculation, this technique may be too slow to be practical. One can still simulate the experiments with a smaller number of events per experiment than were actually obtained. Depending on the estimator it may be possible to construct it for an experiment consisting of a single event. From these one then determines the matrix of correlation coefficients ρ_{ij} for the N estimators $y_1, \ldots y_N$. For efficient estimators (which is the case for maximum likelihood and least squares estimators in the large sample limit), the covariance is inversely proportional to the sample size (cf. equation (6.20)), and thus ρ_{ij} is independent of the number of events in the individual subsamples. One can then use $V_{ij} = \rho_{ij}\sigma_i\sigma_j$, where σ_i and σ_j can be estimated either from the data directly or from the Monte Carlo. If enough data are available, this technique can be applied without recourse to Monte Carlo calculations by dividing the real data sample into a large number of subsamples, determining the estimators $y_1, \ldots, y_N$ for each and from these estimating the matrix of correlation coefficients.

Even if the covariance matrix is not known accurately, the technique is still a valid way to average measurements, and it may represent the best practical solution. Recall that any choice of the weights w_i in equation (7.29), as long as they sum to unity, will lead to an unbiased estimator $\hat{\lambda}$. This still holds if the covariance matrix is only known approximately, but in such a case one will not attain the smallest possible variance and the variance of the average will be incorrectly estimated.

8
The method of moments

Although the methods of maximum likelihood and least squares lead to estimators with optimal or nearly optimal properties, they are sometimes difficult to implement. An alternative technique of parameter estimation is the **method of moments** (MM). Although the variances of MM estimators are not in general as small as those from maximum likelihood, the technique is often simpler from a computational standpoint.

Suppose one has a set of n observations of a random variable x, $x_1, \ldots, x_n$, and a hypothesis for the form of the underlying p.d.f. $f(x; \boldsymbol{\theta})$, which contains m unknown parameters $\boldsymbol{\theta} = (\theta_1, \ldots, \theta_m)$. The idea is to first construct m linearly independent functions $a_i(x)$, $i = 1, \ldots, m$. The $a_i(x)$ are themselves random variables whose expectation values $e_i = E[a_i(x)]$ are functions of the true parameters,

$$E[a_i(x)] = \int a_i(x) f(x; \boldsymbol{\theta}) dx \equiv e_i(\boldsymbol{\theta}). \qquad (8.1)$$

The functions $a_i(x)$ must be chosen such that the expectation values (8.1) can be computed, so that the functions $e_i(\boldsymbol{\theta})$ can be determined.

Since we have seen in Section 5.2 that the sample mean is an unbiased estimator for the expectation value of a random variable, we can estimate the expectation value $e_i = E[a_i(x)]$ by the arithmetic mean of the function $a_i(x)$ evaluated with the observed values of x,

$$\hat{e}_i = \bar{a}_i = \frac{1}{n} \sum_{j=1}^{n} a_i(x_j). \qquad (8.2)$$

The MM estimators for the parameters $\boldsymbol{\theta}$ are defined by setting the expectation values $e_i(\boldsymbol{\theta})$ equal to the corresponding estimators $\hat{e}_i$ and solving for the parameters. That is, one solves the following system of m equations for $\hat{\theta}_1, \ldots, \hat{\theta}_m$:

$$e_1(\hat{\boldsymbol{\theta}}) \;=\; \frac{1}{n}\sum_{i=1}^{n} a_1(x_i)$$

$$\vdots \tag{8.3}$$

$$e_m(\hat{\boldsymbol{\theta}}) \;=\; \frac{1}{n}\sum_{i=1}^{n} a_m(x_i).$$

Possible choices for the functions $a_i(x)$ are integer powers of the variable x: $x^1,\ldots,x^m$, so that the expectation value $E[a_i(x)] = E[x^i]$ is the ith algebraic moment of x (hence the name 'method of moments'). Other sets of m linearly independent functions are possible, however, as long as one can compute their expectation values and obtain m independent functions of the parameters.

We would also like to estimate the covariance matrix for the estimators $\hat{\theta}_1,\ldots,\hat{\theta}_m$. In order to obtain this we first estimate the covariance $\mathrm{cov}[a_i(x),a_j(x)]$ using equation (5.11),

$$\widehat{\mathrm{cov}}[a_i(x),a_j(x)] = \frac{1}{n-1}\sum_{k=1}^{n}(a_i(x_k)-\overline{a}_i)(a_j(x_k)-\overline{a}_j). \tag{8.4}$$

This can be related to the covariance $\mathrm{cov}[\overline{a}_i,\overline{a}_j]$ of the arithmetic means of the functions by

$$\begin{aligned}
\mathrm{cov}[\overline{a}_i,\overline{a}_j] &= \mathrm{cov}\left[\frac{1}{n}\sum_{k=1}^{n}a_i(x_k),\,\frac{1}{n}\sum_{l=1}^{n}a_j(x_l)\right] \\[2mm]
&= \frac{1}{n^2}\sum_{k,l=1}^{n}\mathrm{cov}[a_i(x_k),a_j(x_l)] \\[2mm]
&= \frac{1}{n}\,\mathrm{cov}[a_i,a_j]. \tag{8.5}
\end{aligned}$$

The last line follows from the fact there are n terms in the sum over k and l with $k = l$, which each give $\mathrm{cov}[a_i,a_j]$. The other $n^2 - n$ terms have $k \neq l$, for which the covariance $\mathrm{cov}[a_i(x_k),a_j(x_l)]$ vanishes, since the individual x values are independent. The covariance matrix $\mathrm{cov}[\hat{e}_i,\hat{e}_j]$ for the estimators of the expectation values $\hat{e}_i = \overline{a}_i$ can thus be estimated by

$$\widehat{\mathrm{cov}}[\hat{e}_i,\hat{e}_j] = \frac{1}{n(n-1)}\sum_{k=1}^{n}(a_i(x_k)-\overline{a}_i)\,(a_j(x_k)-\overline{a}_j). \tag{8.6}$$

In order to obtain the covariance matrix $\mathrm{cov}[\hat{\theta}_i,\hat{\theta}_j]$ for the estimators of the parameters themselves, one can then use equation (8.6) with the error propagation formula (1.54),

$$\mathrm{cov}[\hat{\theta}_i, \hat{\theta}_j] = \sum_{k,l} \frac{\partial\hat{\theta}_i}{\partial\hat{e}_k} \frac{\partial\hat{\theta}_j}{\partial\hat{e}_l} \, \mathrm{cov}[\hat{e}_k, \hat{e}_l]. \tag{8.7}$$

Note that even though the value of each measurement x_i is used (i.e. there is no binning of the data) one does not in general exhaust all of the information about the form of the p.d.f. For example, with $a_i(x) = x^i, i = 1, \ldots, m$, only information about the first m moments of x is used, but some of the parameters may be more sensitive to higher moments. For this reason the MM estimators have in general larger variances than those obtained from the principles of maximum likelihood or least squares (cf. [Ead71] Section 8.2.2, [Fro79] Chapters 11 and 12). Because of its simplicity, however, the method of moments is particularly useful if the estimation procedure must be repeated a large number of times.

As an example consider the p.d.f. for the continuous random variable x given by

$$f(x; \alpha, \beta) = \frac{1 + \alpha x + \beta x^2}{d_1 + \alpha d_2 + \beta d_3}, \tag{8.8}$$

with $x_{\min} \le x \le x_{\max}$ and where

$$d_k = \frac{1}{k} \left(x_{\max}^k - x_{\min}^k \right). \tag{8.9}$$

We have already encountered this p.d.f. in Section 6.8, where the parameters α and β were estimated using the method of maximum likelihood; here for comparison we will use the method of moments. For this we need two linearly independent functions of x, which should be chosen such that their expectation values can easily be computed. A rather obvious choice is

$$\begin{aligned} a_1 &= x, \\ a_2 &= x^2. \end{aligned} \tag{8.10}$$

The expectation values $e_1 = E[a_1]$ and $e_2 = E[a_2]$ are found to be

$$\begin{aligned} e_1 &= \frac{d_2 + \alpha d_3 + \beta d_4}{d_1 + \alpha d_2 + \beta d_3}, \\[2mm] e_2 &= \frac{d_3 + \alpha d_4 + \beta d_5}{d_1 + \alpha d_2 + \beta d_3}, \end{aligned} \tag{8.11}$$

with d_n again given by equation (8.9). Solving these two equations for α and β and replacing e_1 and e_2 by $\hat{e}_1$ and $\hat{e}_2$ gives the MM estimators,

$$\hat{\alpha} = \frac{(\hat{e}_1 d_3 - d_4)(\hat{e}_2 d_1 - d_3) - (\hat{e}_1 d_1 - d_2)(\hat{e}_2 d_3 - d_5)}{(\hat{e}_1 d_2 - d_3)(\hat{e}_2 d_3 - d_5) - (\hat{e}_1 d_3 - d_4)(\hat{e}_2 d_2 - d_4)},$$

$$\hat{\beta} = \frac{(\hat{e}_1 d_1 - d_2)(\hat{e}_2 d_2 - d_4) - (\hat{e}_1 d_2 - d_3)(\hat{e}_2 d_1 - d_3)}{(\hat{e}_1 d_2 - d_3)(\hat{e}_2 d_3 - d_5) - (\hat{e}_1 d_3 - d_4)(\hat{e}_2 d_2 - d_4)}.$$

(8.12)

From the example of Section 6.8 we had a data sample of 2000 x values generated with $\alpha = 0.5$, $\beta = 0.5$, $x_{\min} = -0.95$, $x_{\max} = 0.95$. Using the same data here gives

$$\hat{\alpha} = 0.493 \pm 0.051$$
$$\hat{\beta} = 0.410 \pm 0.106.$$

The statistical errors are obtained by means of error propagation from the co-variance matrix for $\hat{e}_1$ and $\hat{e}_2$, which is estimated using equation (8.6). Similarly one obtains the correlation coefficient $r = 0.42$.

These results are similar to those obtained using maximum likelihood, and the estimated standard deviations are actually slightly smaller. The latter fact is, however, the result of a statistical fluctuation in estimating the variance. The true variances of MM estimators are in general greater than or equal to those of the ML estimators. For this particular example they are almost the same. The method of moments has the advantage, however, that the estimates can be obtained without having to maximize the likelihood function, which in this example would require a more complicated numerical calculation.

9

Statistical errors, confidence intervals and limits

In Chapters 5–8, several methods for estimating properties of p.d.f.s (moments and other parameters) have been discussed along with techniques for obtaining the variance of the estimators. Up to now the topic of 'error analysis' has been limited to reporting the variances (and covariances) of estimators, or equivalently the standard deviations and correlation coefficients. This turns out to be inadequate in certain cases, and other ways of communicating the statistical uncertainty of a measurement must be found.

After reviewing in Section 9.1 what is meant by reporting the standard deviation as an estimate of statistical uncertainty, the **confidence interval** is introduced in Section 9.2. This allows for a quantitative statement about the fraction of times that such an interval would contain the true value of the parameter in a large number of repeated experiments. Confidence intervals are treated for a number of important cases in Sections 9.3 through 9.6, and are extended to the multidimensional case in Section 9.7. In Sections 9.8 and 9.9, both Bayesian and classical confidence intervals are used to estimate limits on parameters near a physically excluded region.

9.1 The standard deviation as statistical error

Suppose the result of an experiment is an estimate of a certain parameter. The variance (or equivalently its square root, the standard deviation) of the estimator is a measure of how widely the estimates would be distributed if the experiment were to be repeated many times with the same number of observations per experiment. As such, the standard deviation σ is often reported as the statistical uncertainty of a measurement, and is referred to as the **standard error**.

For example, suppose one has n observations of a random variable x and a hypothesis for the p.d.f. $f(x; \theta)$ which contains an unknown parameter θ. From the sample $x_1, \ldots, x_n$ a function $\hat{\theta}(x_1, \ldots, x_n)$ is constructed (e.g. using maximum likelihood) as an estimator for θ. Using one of the techniques discussed in Chapters 5–8 (e.g. analytic method, RCF bound, Monte Carlo, graphical) the standard deviation of $\hat{\theta}$ can be estimated. Let $\hat{\theta}_{obs}$ be the value of the estimator actually observed, and $\hat{\sigma}_{\hat{\theta}}$ the estimate of its standard deviation. In reporting the measurement of θ as $\hat{\theta}_{obs} \pm \hat{\sigma}_{\hat{\theta}}$ one means that repeated estimates all based on n observations of x would be distributed according to a p.d.f. $g(\hat{\theta})$ centered around some true value θ and true standard deviation $\sigma_{\hat{\theta}}$, which are estimated to be $\hat{\theta}_{\text{obs}}$ and $\hat{\sigma}_{\hat{\theta}}$.

For most practical estimators, the sampling p.d.f. $g(\hat{\theta})$ becomes approximately Gaussian in the large sample limit. If more than one parameter is estimated, then the p.d.f. will become a multidimensional Gaussian characterized by a covariance matrix V. Thus by estimating the standard deviation, or for more than one parameter the covariance matrix, one effectively summarizes all of the information available about how repeated estimates would be distributed. By using the error propagation techniques of Section 1.6, the covariance matrix also gives the equivalent information, at least approximately, for functions of the estimators.

Although the 'standard deviation' definition of statistical error bars could in principle be used regardless of the form of the estimator's p.d.f. $g(\hat{\theta})$, it is not, in fact, the conventional definition if $g(\hat{\theta})$ is not Gaussian. In such cases, one usually reports confidence intervals as described in the next section; this can in general lead to asymmetric error bars. In Section 9.3 it is shown that if $g(\hat{\theta})$ is Gaussian, then the so-called 68.3% confidence interval is the same as the interval covered by $\hat{\theta}_{\text{obs}} \pm \hat{\sigma}_{\hat{\theta}}$.

9.2 Classical confidence intervals (exact method)

An alternative (and often equivalent) method of reporting the statistical error of a measurement is with a confidence interval, which was first developed by Neyman [Ney37]. Suppose as above that one has n observations of a random variable x which can be used to evaluate an estimator $\hat{\theta}(x_1, \ldots, x_n)$ for a parameter θ, and that the value obtained is $\hat{\theta}_{\text{obs}}$. Furthermore, suppose that by means of, say, an analytical calculation or a Monte Carlo study, one knows the p.d.f. of $\hat{\theta}$, $g(\hat{\theta}; \theta)$, which contains the true value θ as a parameter. That is, the real value of θ is not known, but for a given θ, one knows what the p.d.f. of $\hat{\theta}$ would be.

Figure 9.1 shows a probability density for an estimator $\hat{\theta}$ for a particular value of the true parameter θ. From $g(\hat{\theta}; \theta)$ one can determine the value u_α such that there is a fixed probability α to observe $\hat{\theta} \geq u_\alpha$, and similarly the value v_β such that there is a probability β to observe $\hat{\theta} \leq v_\beta$. The values u_α and v_β depend on the true value of θ, and are thus determined by

$$\alpha = P(\hat{\theta} \geq u_\alpha(\theta)) = \int_{u_\alpha(\theta)}^{\infty} g(\hat{\theta}; \theta) d\hat{\theta} = 1 - G(u_\alpha(\theta); \theta), \tag{9.1}$$

and

$$\beta = P(\hat{\theta} \leq v_\beta(\theta)) = \int_{-\infty}^{v_\beta(\theta)} g(\hat{\theta}; \theta) d\hat{\theta} = G(v_\beta(\theta); \theta), \tag{9.2}$$

where G is the cumulative distribution corresponding to the p.d.f. $g(\hat{\theta}; \theta)$.

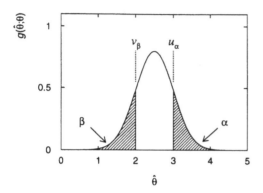

Fig. 9.1 A p.d.f. $g(\hat{\theta}; \theta)$ for an estimator $\hat{\theta}$ for a given value of the true parameter θ. The two shaded regions indicate the values of $\hat{\theta} \leq v_\beta$, which has a probability β, and $\hat{\theta} \geq u_\alpha$, which has a probability α.

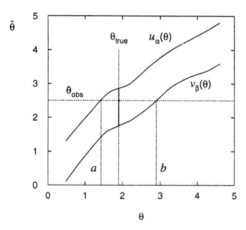

Fig. 9.2 Construction of the confidence interval $[a, b]$ given an observed value $\hat{\theta}_{obs}$ of the estimator $\hat{\theta}$ for the parameter θ (see text).

Figure 9.2 shows an example of how the functions $u_\alpha(\theta)$ and $v_\beta(\theta)$ might appear as a function of the true value of θ. The region between the two curves is called the **confidence belt**. The probability for the estimator to be inside the belt, regardless of the value of θ, is given by

$$P(v_\beta(\theta) \leq \hat{\theta} \leq u_\alpha(\theta)) = 1 - \alpha - \beta. \tag{9.3}$$

As long as $u_\alpha(\theta)$ and $v_\beta(\theta)$ are monotonically increasing functions of θ, which in general should be the case if $\hat{\theta}$ is to be a good estimator for θ, one can determine the inverse functions

$$a(\hat{\theta}) \equiv u_\alpha^{-1}(\hat{\theta}),$$
$$b(\hat{\theta}) \equiv v_\beta^{-1}(\hat{\theta}). \tag{9.4}$$

The inequalities

$$\hat{\theta} \geq u_\alpha(\theta),$$
$$\hat{\theta} \leq v_\beta(\theta),$$

(9.5)

then imply respectively

$$a(\hat{\theta}) \geq \theta,$$
$$b(\hat{\theta}) \leq \theta.$$

(9.6)

Equations (9.1) and (9.2) thus become

$$P(a(\hat{\theta}) \geq \theta) = \alpha,$$
$$P(b(\hat{\theta}) \leq \theta) = \beta,$$

(9.7)

or taken together,

$$P(a(\hat{\theta}) \leq \theta \leq b(\hat{\theta})) = 1 - \alpha - \beta.$$

(9.8)

If the functions $a(\hat{\theta})$ and $b(\hat{\theta})$ are evaluated with the value of the estimator actually obtained in the experiment, $\hat{\theta}_{\mathrm{obs}}$, then this determines two values, a and b, as illustrated in Fig. 9.2. The interval $[a, b]$ is called a **confidence interval** at a **confidence level** or **coverage probability** of $1 - \alpha - \beta$. The idea behind its construction is that the coverage probability expressed by equations (9.7), and hence also (9.8), holds regardless of the true value of θ, which of course is unknown. It should be emphasized that a and b are random values, since they depend on the estimator $\hat{\theta}$, which is itself a function of the data. If the experiment were repeated many times, the interval $[a, b]$ would include the true value of the parameter θ in a fraction $1 - \alpha - \beta$ of the experiments.

The relationship between the interval $[a, b]$ and its coverage probability $1 - \alpha - \beta$ can be understood from Fig. 9.2 by considering the hypothetical true value indicated as θ_{true}. If this is the true value of θ, then $\hat{\theta}_{\mathrm{obs}}$ will intersect the solid segment of the vertical line between $u_\alpha(\theta_{\mathrm{true}})$ and $v_\beta(\theta_{\mathrm{true}})$ with a probability of $1 - \alpha - \beta$. From the figure one can see that the interval $[a, b]$ will cover θ_{true} if $\hat{\theta}_{\mathrm{obs}}$ intersects this segment, and will not otherwise.

In some situations one may only be interested in a **one-sided confidence interval** or **limit**. That is, the value a represents a lower limit on the parameter θ such that $a \leq \theta$ with the probability $1 - \alpha$. Similarly, b represents an upper limit on θ such that $P(\theta \leq b) = 1 - \beta$.

Two-sided intervals (i.e. both a and b specified) are not uniquely determined by the confidence level $1 - \alpha - \beta$. One often chooses, for example, $\alpha = \beta = \gamma/2$ giving a so-called **central confidence interval** with probability $1 - \gamma$. Note that a central confidence interval does not necessarily mean that a and b are equidistant from the estimated value $\hat{\theta}$, but only that the probabilities α and β are equal.

By construction, the value a gives the hypothetical value of the true parameter θ for which a fraction α of repeated estimates $\hat{\theta}$ would be higher than the one actually obtained, $\hat{\theta}_{obs}$, as is illustrated in Fig. 9.3. Similarly, b is the value of θ for which a fraction β of the estimates would be lower than $\hat{\theta}_{obs}$. That is, taking $\hat{\theta}_{obs} = u_\alpha(a) = v_\beta(b)$, equations (9.1) and (9.2) become

$$\alpha = \int_{\hat{\theta}_{obs}}^{\infty} g(\hat{\theta}; a)\, d\hat{\theta} = 1 - G(\hat{\theta}_{obs}; a),$$

$$\beta = \int_{-\infty}^{\hat{\theta}_{obs}} g(\hat{\theta}; b)\, d\hat{\theta} = G(\hat{\theta}_{obs}; b). \tag{9.9}$$

The previously described procedure to determine the confidence interval is thus equivalent to solving (9.9) for a and b, e.g. numerically.

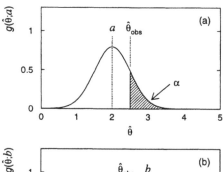

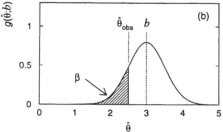

Fig. 9.3 (a) The p.d.f. $g(\hat{\theta}; a)$, where a is the lower limit of the confidence interval. If the true parameter θ were equal to a, the estimates $\hat{\theta}$ would be greater than the one actually observed $\hat{\theta}_{obs}$ with a probability α. (b) The p.d.f. $g(\hat{\theta}; b)$, where b is the upper limit of the confidence interval. If θ were equal to b, $\hat{\theta}$ would be observed less than $\hat{\theta}_{obs}$ with probability β.

Figure 9.3 also illustrates the relationship between a confidence interval and a test of goodness-of-fit, cf. Section 4.5. For example, we could test the hypothesis $\theta = a$ using $\hat{\theta}$ as a test statistic. If we define the region $\hat{\theta} \geq \hat{\theta}_{obs}$ as having equal or less agreement with the hypothesis than the result obtained (a one-sided test), then the resulting P-value of the test is α. For the confidence interval, however, the probability α is specified first, and the value a is a random quantity depending on the data. For a goodness-of-fit test, the hypothesis, here $\theta = a$, is specified and the P-value is treated as a random variable.

Note that one sometimes calls the P-value, here equal to α, the 'confidence level' of the test, whereas the one-sided confidence interval $\theta \geq a$ has a confidence level of $1 - \alpha$. That is, for a test, small α indicates a low level of confidence in the hypothesis $\theta = a$. For a confidence interval, small α indicates a *high* level of confidence that the interval $\theta \geq a$ includes the true parameter. To avoid confusion we will use the term P-value or (observed) significance level for goodness-of-fit tests, and reserve the term confidence level to mean the coverage probability of a confidence interval.

The confidence interval $[a, b]$ is often expressed by reporting the result of a measurement as $\hat{\theta}_{-c}^{+d}$, where $\hat{\theta}$ is the estimated value, and $c = \hat{\theta} - a$ and $d = b - \hat{\theta}$ are usually displayed as **error bars**. In many cases the p.d.f. $g(\hat{\theta}; \theta)$ is approximately Gaussian, so that an interval of plus or minus one standard deviation around the measured value corresponds to a central confidence interval with $1 - \gamma = 0.683$ (see Section 9.3). The 68.3% central confidence interval is usually adopted as the conventional definition for error bars even when the p.d.f. of the estimator is not Gaussian.

If, for example, the result of an experiment is reported as $\hat{\theta}_{-c}^{+d} = 5.79_{-0.25}^{+0.32}$, it is meant that if one were to construct the interval $[\hat{\theta} - c, \hat{\theta} + d]$ according to the prescription described above in a large number of similar experiments with the same number of measurements per experiment, then the interval would include the true value θ in $1 - \alpha - \beta$ of the cases. It does not mean that the probability (in the sense of relative frequency) that the true value of θ is in the fixed interval $[5.54, 6.11]$ is $1 - \alpha - \beta$. In the frequency interpretation, the true parameter θ is not a random variable and is assumed to not fluctuate from experiment to experiment. In this sense the probability that θ is in $[5.54, 6.11]$ is either 0 or 1, but we do not know which. The interval itself, however, is subject to fluctuations since it is constructed from the data.

A difficulty in constructing confidence intervals is that the p.d.f. of the estimator $g(\hat{\theta}; \theta)$, or equivalently the cumulative distribution $G(\hat{\theta}; \theta)$, must be known. An example is given in Section 10.4, where the p.d.f. for the estimator of the mean ξ of an exponential distribution is derived, and from this a confidence interval for ξ is determined. In many practical applications, estimators are Gaussian distributed (at least approximately). In this case the confidence interval can be determined easily; this is treated in detail in the next section. Even in the case of a non-Gaussian estimator, however, a simple approximate technique can be applied using the likelihood function; this is described in Section 9.6.

9.3 Confidence interval for a Gaussian distributed estimator

A simple and very important application of a confidence interval is when the distribution of $\hat{\theta}$ is Gaussian with mean θ and standard deviation $\sigma_{\hat{\theta}}$. That is, the cumulative distribution of $\hat{\theta}$ is

$$G(\hat{\theta}; \theta, \sigma_{\hat{\theta}}) = \int_{-\infty}^{\hat{\theta}} \frac{1}{\sqrt{2\pi\sigma_{\hat{\theta}}^2}} \exp\left(\frac{-(\hat{\theta}' - \theta)^2}{2\sigma_{\hat{\theta}}^2}\right) d\hat{\theta}'. \tag{9.10}$$

This is a commonly occurring situation since, according to the central limit theorem, any estimator that is a linear function of a sum of random variables becomes Gaussian in the large sample limit. We will see that for this case, the

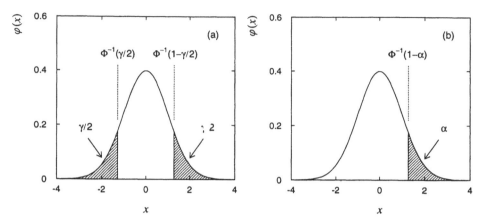

Fig. 9.4 The standard Gaussian p.d.f. $\varphi(x)$ showing the relationship between the quantiles Φ^{-1} and the confidence level for (a) a central confidence interval and (b) a one-sided confidence interval.

somewhat complicated procedure explained in the previous section results in a simple prescription for determining the confidence interval.

Suppose that the standard deviation $\sigma_{\hat{\theta}}$ is known, and that the experiment has resulted in an estimate $\hat{\theta}_{\text{obs}}$. According to equations (9.9), the confidence interval $[a, b]$ is determined by solving the equations

$$\alpha = 1 - G(\hat{\theta}_{\text{obs}}; a, \sigma_{\hat{\theta}}) = 1 - \Phi\left(\frac{\hat{\theta}_{\text{obs}} - a}{\sigma_{\hat{\theta}}}\right),$$

$$\beta = G(\hat{\theta}_{\text{obs}}; b, \sigma_{\hat{\theta}}) = \Phi\left(\frac{\hat{\theta}_{\text{obs}} - b}{\sigma_{\hat{\theta}}}\right),$$

(9.11)

for a and b, where G has been expressed using the cumulative distribution of the standard Gaussian Φ (2.26) (see also (2.27)). This gives

$$a = \hat{\theta}_{\text{obs}} - \sigma_{\hat{\theta}}\, \Phi^{-1}(1 - \alpha),$$

$$b = \hat{\theta}_{\text{obs}} + \sigma_{\hat{\theta}}\, \Phi^{-1}(1 - \beta).$$

(9.12)

Here Φ^{-1} is the inverse function of Φ, i.e. the quantile of the standard Gaussian, and in order to make the two equations symmetric we have used $\Phi^{-1}(\beta) = -\Phi^{-1}(1 - \beta)$.

The quantiles $\Phi^{-1}(1 - \alpha)$ and $\Phi^{-1}(1 - \beta)$ represent how far away the interval limits a and b are located with respect to the estimate $\hat{\theta}_{\text{obs}}$ in units of the standard deviation $\sigma_{\hat{\theta}}$. The relationship between the quantiles of the standard Gaussian distribution and the confidence level is illustrated in Fig. 9.4(a) for central and Fig. 9.4(b) for one-sided confidence intervals.

Consider a central confidence interval with $\alpha = \beta = \gamma/2$. The confidence level $1 - \gamma$ is often chosen such that the quantile is a small integer, e.g. $\Phi^{-1}(1 - \gamma/2) = 1, 2, 3, \ldots$. Similarly, for one-sided intervals (limits) one often chooses a small integer for $\Phi^{-1}(1 - \alpha)$. Commonly used values for both central and one-sided intervals are shown in Table 9.1. Alternatively one can choose a round number for the confidence level instead of for the quantile. Commonly used values are shown in Table 9.2. Other possible values can be obtained from [Bra92, Fro79, Dud88] or from computer routines (e.g. the routine `GAUSIN` in [CER97]).

Table 9.1 The values of the confidence level for different values of the quantile of the standard Gaussian Φ^{-1}: for central intervals (left) the quantile $\Phi^{-1}(1 - \gamma/2)$ and confidence level $1 - \gamma$; for one-sided intervals (right) the quantile $\Phi^{-1}(1 - \alpha)$ and confidence level $1 - \alpha$.

$\Phi^{-1}(1 - \gamma/2)$	$1 - \gamma$	$\Phi^{-1}(1 - \alpha)$	$1 - \alpha$
1	0.6827	1	0.8413
2	0.9544	2	0.9772
3	0.9973	3	0.9987

Table 9.2 The values of the quantile of the standard Gaussian Φ^{-1} for different values of the confidence level: for central intervals (left) the confidence level $1 - \gamma$ and the quantile $\Phi^{-1}(1 - \gamma/2)$; for one-sided intervals (right) the confidence level $1 - \alpha$ and the quantile $\Phi^{-1}(1 - \alpha)$.

$1 - \gamma$	$\Phi^{-1}(1 - \gamma/2)$	$1 - \alpha$	$\Phi^{-1}(1 - \alpha)$
0.90	1.645	0.90	1.282
0.95	1.960	0.95	1.645
0.99	2.576	0.99	2.326

For the conventional 68.3% central confidence interval one has $\alpha = \beta = \gamma/2$, with $\Phi^{-1}(1 - \gamma/2) = 1$, i.e. a '1 σ error bar'. This results in the simple prescription

$$[a, b] = [\hat{\theta}_{\text{obs}} - \sigma_{\hat{\theta}}, \hat{\theta}_{\text{obs}} + \sigma_{\hat{\theta}}]. \tag{9.13}$$

Thus for the case of a Gaussian distributed estimator, the 68.3% central confidence interval is given by the estimated value plus or minus one standard deviation. The final result of the measurement of θ is then simply reported as $\hat{\theta}_{\text{obs}} \pm \sigma_{\hat{\theta}}$.

If the standard deviation $\sigma_{\hat{\theta}}$ is not known a priori but rather is estimated from the data, then the situation is in principle somewhat more complicated. If, for example, the estimated standard deviation $\hat{\sigma}_{\hat{\theta}}$ had been used instead of $\sigma_{\hat{\theta}}$, then it would not have been so simple to relate the cumulative distribution $G(\hat{\theta}; \theta, \hat{\sigma}_{\hat{\theta}})$ to Φ, the cumulative distribution of the standard Gaussian, since $\hat{\sigma}_{\hat{\theta}}$ depends in general on $\hat{\theta}$. In practice, however, the recipe given above can still be applied using the estimate $\hat{\sigma}_{\hat{\theta}}$ instead of $\sigma_{\hat{\theta}}$, as long as $\hat{\sigma}_{\hat{\theta}}$ is a sufficiently good approximation of the true standard deviation, e.g. for a large enough data sample. For the small sample case where $\hat{\theta}$ represents the mean of n Gaussian random variables of unknown standard deviation, the confidence interval can be determined by relating the cumulative distribution $G(\hat{\theta}; \theta, \hat{\sigma}_{\hat{\theta}})$ to Student's t distribution (see e.g. [Fro79], [Dud88] Section 10.2).

Exact determination of confidence intervals becomes more difficult if the p.d.f. of the estimator $g(\hat{\theta}; \theta)$ is not Gaussian, or worse, if it is not known analytically. For a non-Gaussian p.d.f. it is sometimes possible to transform the parameter $\theta \to \eta(\theta)$ such that the p.d.f. for the estimator $\hat{\eta}$ is approximately Gaussian. The confidence interval for the transformed parameter η can then be converted back into an interval for θ. An example of this technique is given in Section 9.5.

9.4 Confidence interval for the mean of the Poisson distribution

Along with the Gaussian distributed estimator, another commonly occurring case is where the outcome of a measurement is a Poisson variable n ($n = 0, 1, 2, \ldots$). Recall from (2.9) that the probability to observe n is

$$f(n; \nu) = \frac{\nu^n}{n!} e^{-\nu},\qquad(9.14)$$

and that the parameter ν is equal to the expectation value $E[n]$. The maximum likelihood estimator for ν can easily be found to be $\hat{\nu} = n$. Suppose that a single measurement has resulted in the value $\hat{\nu}_{\text{obs}} = n_{\text{obs}}$, and that from this we would like to construct a confidence interval for the mean ν.

For the case of a discrete variable, the procedure for determining the confidence interval described in Section 9.2 cannot be directly applied. This is because the functions $u_\alpha(\theta)$ and $v_\beta(\theta)$, which determine the confidence belt, do not exist for all values of the parameter θ. For the Poisson case, for example, we would need to find $u_\alpha(\nu)$ and $v_\beta(\nu)$ such that $P(\hat{\nu} \geq u_\alpha(\nu)) = \alpha$ and $P(\hat{\nu} \leq v_\beta(\nu)) = \beta$ for all values of the parameter ν. But if α and β are fixed, then because $\hat{\nu}$ only takes on discrete values, these equations hold in general only for particular values of ν.

A confidence interval $[a, b]$ can still be determined, however, by using equations (9.9). For the case of a discrete random variable and a parameter ν these become

$$\begin{aligned}\alpha &= P(\hat{\nu} \geq \hat{\nu}_{\text{obs}}; a), \\ \beta &= P(\hat{\nu} \leq \hat{\nu}_{\text{obs}}; b),\end{aligned}\qquad(9.15)$$

and in particular for a Poisson variable one has

$$\alpha = \sum_{n=n_{\text{obs}}}^{\infty} f(n; a) = 1 - \sum_{n=0}^{n_{\text{obs}}-1} f(n; a) = 1 - \sum_{n=0}^{n_{\text{obs}}-1} \frac{a^n}{n!} e^{-a},$$

$$\beta = \sum_{n=0}^{n_{\text{obs}}} f(n; b) = \sum_{n=0}^{n_{\text{obs}}} \frac{b^n}{n!} e^{-b}. \tag{9.16}$$

For an estimate $\hat{\nu} = n_{\text{obs}}$ and given probabilities α and β, these equations can be solved numerically for a and b. Here one can use the following relation between the Poisson and χ^2 distributions,

$$\sum_{n=0}^{n_{\text{obs}}} \frac{\nu^n}{n!} e^{-\nu} = \int_{2\nu}^{\infty} f_{\chi^2}(z; n_{\text{d}} = 2(n_{\text{obs}} + 1)) \, dz$$

$$= 1 - F_{\chi^2}(2\nu; n_{\text{d}} = 2(n_{\text{obs}} + 1)), \tag{9.17}$$

where f_{χ^2} is the χ^2 p.d.f. for n_{d} degrees of freedom and F_{χ^2} is the corresponding cumulative distribution. One then has

$$a = \tfrac{1}{2} F_{\chi^2}^{-1}(\alpha; n_{\text{d}} = 2n_{\text{obs}}),$$

$$b = \tfrac{1}{2} F_{\chi^2}^{-1}(1 - \beta; n_{\text{d}} = 2(n_{\text{obs}} + 1)). \tag{9.18}$$

Quantiles $F_{\chi^2}^{-1}$ of the χ^2 distribution can be obtained from standard tables (e.g. in [Bra92]) or from computer routines such as CHISIN in [CER97]. Some values for $n_{\text{obs}} = 0, \ldots, 10$ are shown in Table 9.3.

Note that the lower limit a cannot be determined if $n_{\text{obs}} = 0$. Equations (9.15) say that if $\nu = a$ ($\nu = b$), then the probability is α (β) to observe a value greater (less) than or equal to the one actually observed. Because the case of equality, $\hat{\nu} = \hat{\nu}_{\text{obs}}$, is included in the inequalities (9.15), one obtains a conservatively large confidence interval, i.e.

$$P(\nu \geq a) \geq 1 - \alpha,$$

$$P(\nu \leq b) \geq 1 - \beta, \tag{9.19}$$

$$P(a \leq \nu \leq b) \geq 1 - \alpha - \beta.$$

An important special case is when the observed number n_{obs} is zero, and one is interested in establishing an upper limit b. Equation (9.15) becomes

$$\beta = \sum_{n=0}^{0} \frac{b^n e^{-b}}{n!} = e^{-b}, \tag{9.20}$$

Table 9.3 Poisson lower and upper limits for n_{obs} observed events.

n_{obs}	lower limit a			upper limit b		
	$\alpha = 0.1$	$\alpha = 0.05$	$\alpha = 0.01$	$\beta = 0.1$	$\beta = 0.05$	$\beta = 0.01$
0	–	–	–	2.30	3.00	4.61
1	0.105	0.051	0.010	3.89	4.74	6.64
2	0.532	0.355	0.149	5.32	6.30	8.41
3	1.10	0.818	0.436	6.68	7.75	10.04
4	1.74	1.37	0.823	7.99	9.15	11.60
5	2.43	1.97	1.28	9.27	10.51	13.11
6	3.15	2.61	1.79	10.53	11.84	14.57
7	3.89	3.29	2.33	11.77	13.15	16.00
8	4.66	3.98	2.91	12.99	14.43	17.40
9	5.43	4.70	3.51	14.21	15.71	18.78
10	6.22	5.43	4.13	15.41	16.96	20.14

or $b = -\log \beta$. For the upper limit at a confidence level of $1 - \beta = 95\%$ one has $b = -\log(0.05) = 2.996 \approx 3$. Thus if the number of occurrences of some rare event is treated as a Poisson variable with mean ν, and one looks for events of this type and finds none, then the 95% upper limit on the mean is 3. That is, if the mean were in fact $\nu = 3$, then the probability to observe zero would be 5%.

9.5 Confidence interval for correlation coefficient, transformation of parameters

In many situations one can assume that the p.d.f. for an estimator is Gaussian, and the results of Section 9.3 can then be used to obtain a confidence interval. As an example where this may not be the case, consider the correlation coefficient ρ of two continuous random variables x and y distributed according to a two-dimensional Gaussian p.d.f. $f(x, y)$ (equation (2.30)). Suppose we have a sample of n independent observations of x and y, and we would like to determine a confidence interval for ρ based on the estimator r, cf. equation (5.12),

$$r = \frac{\sum_{i=1}^{n}(x_i - \overline{x})(y_i - \overline{y})}{\left(\sum_{j=1}^{n}(x_j - \overline{x})^2 \cdot \sum_{k=1}^{n}(y_k - \overline{y})^2\right)^{1/2}}. \tag{9.21}$$

The p.d.f. $g(r; \rho, n)$ has a rather complicated form; it is given, for example, in [Mui82] p. 151. A graph is shown in Fig. 9.5 for a sample of size $n = 20$ for several values of the true correlation coefficient ρ. One can see that $g(r; \rho, n)$ is asymmetric and that the degree of asymmetry depends on ρ. It can be shown that $g(r; \rho, n)$ approaches a Gaussian in the large sample limit, but for this approximation to be valid, one requires a fairly large sample. (At least $n \geq 500$ is recommended [Bra92].) For smaller samples such as in Fig. 9.5, one cannot rely on the Gaussian approximation for $g(r; \rho, n)$, and thus one cannot use (9.12) to determine the confidence interval.

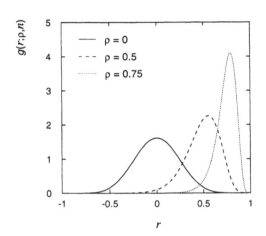

Fig. 9.5 The probability density $f(r; \rho, n)$ for the estimator r of the correlation coefficient ρ shown for a sample of size $n = 20$ and various values of ρ.

In principle one is then forced to return to the procedure of Section 9.2, which in this case would be difficult computationally. There exists, however, a simpler method to determine an approximate confidence interval for ρ. It has been shown by Fisher that the p.d.f. of the statistic

$$z = \tanh^{-1} r = \tfrac{1}{2} \log \frac{1+r}{1-r} \tag{9.22}$$

approaches the Gaussian limit much more quickly as a function of the sample size n than that of r (see [Fis90] and references therein). This can be used as an estimator for ζ, defined as

$$\zeta = \tanh^{-1} \rho = \tfrac{1}{2} \log \frac{1+\rho}{1-\rho}. \tag{9.23}$$

One can show that the expectation value of z is approximately given by

$$E[z] = \tfrac{1}{2} \log \frac{1+\rho}{1-\rho} + \frac{\rho}{2(n-1)} \tag{9.24}$$

and its variance by

$$V[z] = \frac{1}{n-3}. \tag{9.25}$$

We will assume that the sample is large enough that z has a Gaussian p.d.f. and that the bias term $\rho/2(n-1)$ in (9.24) can be neglected. Given a sample of n measurements of x and y, z can be determined according to equation (9.22) and its standard deviation $\hat{\sigma}_z$ can be estimated by using the variance from equation (9.25). One can use these to determine the interval $[z - \hat{\sigma}_z, z + \hat{\sigma}_z]$, or in general the interval $[a, b]$ given by (9.12). These give the lower limit a for ζ with confidence level $1 - \alpha$ and an upper limit b with confidence level $1 - \beta$. The confidence interval $[a, b]$ for $\zeta = \tanh^{-1} \rho$ can then be converted back to an interval $[A, B]$

for ρ simply by using the inverse of the transformation (9.22), i.e. $A = \tanh a$ and $B = \tanh b$.

Consider for example a sample of size $n = 20$ for which one has obtained the estimate $r = 0.5$. From equation (5.17) the standard deviation of r can be estimated as $\hat{\sigma}_r = (1 - r^2)/\sqrt{n} = 0.168$. If one were to make the incorrect approximation that r is Gaussian distributed for such a small sample, this would lead to a 68.3% central confidence interval for ρ of $[0.332, 0.668]$, or $[0.067, 0.933]$ at a confidence level of 99%. Thus since the sample correlation coefficient r is almost three times the standard error $\hat{\sigma}_r$, one might be led to the incorrect conclusion that there is significant evidence for a non-zero value of ρ, i.e. a '3σ effect'. By using the z-transformation, however, one obtains $z = 0.549$ and $\hat{\sigma}_z = 0.243$. This corresponds to a 99% central confidence interval of $[-0.075, 1.174]$ for ζ, and $[-0.075, 0.826]$ for ρ. Thus the 99% central confidence interval includes zero.

Recall that the lower limit of the confidence interval is equal to the hypothetical value of the true parameter such that r would be observed higher than the one actually observed with the probability α. One can ask, for example, what the confidence level would be for a lower limit of zero. If we had assumed that $g(r; \rho, n)$ was Gaussian, the corresponding probability would be 0.14%. By using the z-transformation, however, the confidence level for a limit of zero is 2.3%, i.e. if ρ were zero one would obtain r greater than or equal to the one observed, $r = 0.5$, with a probability of 2.3%. The actual evidence for a non-zero correlation is therefore not nearly as strong as one would have concluded by simply using the standard error $\hat{\sigma}_r$ with the assumption that r is Gaussian.

9.6   Confidence intervals using the likelihood function or χ^2

Even in the case of a non-Gaussian estimator, the confidence interval can be determined with a simple approximate technique which makes use of the likelihood function or equivalently the χ^2 function where one has $L = \exp(-\chi^2/2)$. Consider first a maximum likelihood estimator $\hat{\theta}$ for a parameter θ in the large sample limit. In this limit it can be shown ([Stu91] Chapter 18) that the p.d.f. $g(\hat{\theta}; \theta)$ becomes Gaussian,

$$g(\hat{\theta}; \theta) = \frac{1}{\sqrt{2\pi\sigma_{\hat{\theta}}^2}} \exp\left(\frac{-(\hat{\theta} - \theta)^2}{2\sigma_{\hat{\theta}}^2}\right), \tag{9.26}$$

centered about the true value of the parameter θ and with a standard deviation $\sigma_{\hat{\theta}}$.

One can also show that in the large sample limit the likelihood function itself becomes Gaussian in form centered about the ML estimate $\hat{\theta}$,

$$L(\theta) = L_{\max} \exp\left(\frac{-(\theta - \hat{\theta})^2}{2\sigma_{\hat{\theta}}^2}\right). \tag{9.27}$$

From the RCF inequality (6.16), which for an ML estimator in the large sample limit becomes an equality, one obtains that $\sigma_{\hat{\theta}}$ in the likelihood function (9.27) is the same as in the p.d.f. (9.26). This has already been encountered in Section 6.7, equation (6.24), where the likelihood function was used to estimate the variance of an estimator $\hat{\theta}$. This led to a simple prescription for estimating $\sigma_{\hat{\theta}}$, since by changing the parameter θ by N standard deviations, the log-likelihood function decreases by $N^2/2$ from its maximum value,

$$\log L(\hat{\theta} \pm N\sigma_{\hat{\theta}}) = \log L_{\max} - \frac{N^2}{2}. \tag{9.28}$$

From the results of the previous section, however, we know that for a Gaussian distributed estimator $\hat{\theta}$, the 68.3% central confidence interval can be constructed from the estimator and its estimated standard deviation $\hat{\sigma}_{\hat{\theta}}$ as $[a, b] = [\theta - \sigma_{\hat{\theta}}, \theta + \hat{\sigma}_{\hat{\theta}}]$ (or more generally according to (9.12) for a confidence level of $1 - \gamma$). The 68.3% central confidence interval is thus given by the values of θ at which the log-likelihood function decreases by $1/2$ from its maximum value. (This is assuming, of course, that $\hat{\theta}$ is the ML estimator and thus corresponds to the maximum of the likelihood function.)

In fact, it can be shown that even if the likelihood function is not a Gaussian function of the parameters, the central confidence interval $[a, b] = [\theta - c, \theta + d]$ can still be approximated by using

$$\log L(\hat{\theta}_{-c}^{+d}) = \log L_{\max} - \frac{N^2}{2}, \tag{9.29}$$

where $N = \Phi^{-1}(1 - \gamma/2)$ is the quantile of the standard Gaussian corresponding to the desired confidence level $1 - \gamma$. (For example, $N = 1$ for a 68.3% central confidence interval; see Table 9.1.) In the case of a least squares fit with Gaussian errors, i.e. with $\log L = -\chi^2/2$, the prescription becomes

$$\chi^2(\hat{\theta}_{-c}^{+d}) = \chi^2_{\min} + N^2. \tag{9.30}$$

A heuristic proof that the intervals defined by equations (9.29) and (9.30) approximate the classical confidence intervals of Section 9.2 can be found in [Ead71, Fro79]. Equations (9.29) and (9.30) represent one of the most commonly used methods for estimating statistical uncertainties. One should keep in mind, however, that the correspondence with the method of Section 9.2 is only exact in the large sample limit. Several authors have recommended using the term 'likelihood interval' for an interval obtained from the likelihood function [Fro79, Hud64]. Regardless of the name, it should be kept in mind that it is interpreted here as an approximation to the classical confidence interval, i.e. a random interval constructed so as to include the true parameter value with a given probability.

As an example consider the estimator $\hat{\tau} = \frac{1}{n}\sum_{i=1}^{n} t_i$ for the parameter τ of an exponential distribution, as in the example of Section 6.2 (see also Section 6.7). There, the ML method was used to estimate τ given a sample of $n = 50$ measurements of an exponentially distributed random variable t. This sample was sufficiently large that the standard deviation $\sigma_{\hat{\tau}}$ could be approximated by the values of τ where the log-likelihood function decreased by 1/2 from its maximum (see Fig. 6.4). This gave $\hat{\tau} = 1.06$ and $\hat{\sigma}_{\hat{\tau}} \approx \Delta\hat{\tau}_{-} \approx \Delta\hat{\tau}_{+} \approx 0.15$.

Figure 9.6 shows the log-likelihood function $\log L(\tau)$ as a function of τ for a sample of only $n = 5$ measurements of an exponentially distributed random variable, generated using the Monte Carlo method with the true parameter $\tau = 1$. Because of the smaller sample size the log-likelihood function is less parabolic than before.

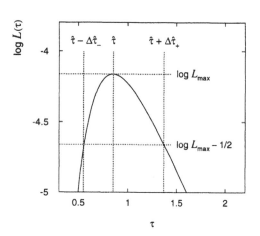

Fig. 9.6 The log-likelihood function $\log L(\tau)$ as a function of τ for a sample of $n = 5$ measurements. The interval $[\hat{\tau} - \Delta\hat{\tau}_{-}, \hat{\tau} + \Delta\hat{\tau}_{+}]$ determined by $\log L(\tau) = \log L_{\max} - 1/2$ can be used to approximate the 68.3% central confidence interval.

One could still use the half-width of the interval determined by $\log L_{\max} - 1/2$ to approximate the standard deviation $\sigma_{\hat{\tau}}$, but this is not really what we want. The statistical uncertainty is better communicated by giving the confidence interval, since one then knows the probability that the interval covers the true parameter value. Furthermore, by giving a central confidence interval (and hence asymmetric errors, $\Delta\hat{\tau}_{-} \neq \Delta\hat{\tau}_{+}$), one has equal probabilities for the true parameter to be higher or lower than the interval limits. As illustrated in Fig. 9.6, the central confidence interval can be approximated by the values of τ where $\log L(\tau) = \log L_{\max} - 1/2$, which gives $[\hat{\tau} - \Delta\hat{\tau}_{-}, \hat{\tau} + \Delta\hat{\tau}_{+}] = [0.55, 1.37]$ or $\hat{\tau} = 0.85^{+0.52}_{-0.30}$.

In fact, the same could have been done in Section 6.7 by giving the result there as $\hat{\tau} = 1.062^{+0.165}_{-0.137}$. Whether one chooses this method or simply reports an averaged symmetric error (i.e. $\hat{\tau} = 1.06 \pm 0.15$) will depend on how accurately the statistical error needs to be given. For the case of $n = 5$ shown in Fig. 9.6, the error bars are sufficiently asymmetric that one would probably want to use the 68.3% central confidence interval and give the result as $\hat{\tau} = 0.85^{+0.52}_{-0.30}$.

9.7 Multidimensional confidence regions

In Section 9.2, a confidence interval $[a, b]$ was constructed so as to have a certain probability $1 - \gamma$ of containing a parameter θ. In order to generalize this

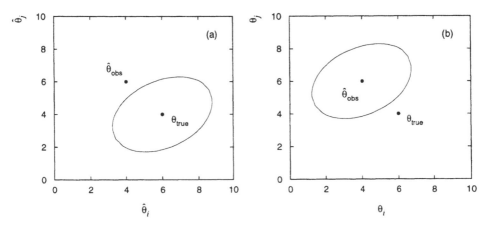

Fig. 9.7 (a) A contour of constant $g(\hat{\theta};\theta_{\text{true}})$ (i.e. constant $Q(\hat{\theta},\theta_{\text{true}})$) in $\hat{\theta}$-space. (b) A contour of constant $L(\theta)$ corresponding to constant $Q(\hat{\theta}_{\text{obs}},\theta)$ in θ-space. The values θ_{true} and $\hat{\theta}_{\text{obs}}$ represent particular constant values of θ and $\hat{\theta}$, respectively.

to the case of n parameters, $\theta = (\theta_1, \ldots, \theta_n)$, one might attempt to find an n-dimensional confidence interval $[\mathbf{a}, \mathbf{b}]$ constructed so as to have a given probability that $a_i < \theta_i < b_i$, simultaneously for all i. This turns out to be computationally difficult, and is rarely done.

It is nevertheless quite simple to construct a **confidence region** in the parameter space such that the true parameter θ is contained within the region with a given probability (at least approximately). This region will not have the form $a_i < \theta_i < b_i$, $i = 1, \ldots, n$, but will be more complicated, approaching an n-dimensional hyperellipsoid in the large sample limit.

As in the single-parameter case, one makes use of the fact that both the joint p.d.f. for the estimator $\hat{\theta} = (\hat{\theta}_1, \ldots, \hat{\theta}_n)$ as well as the likelihood function become Gaussian in the large sample limit. That is, the joint p.d.f. of $\hat{\theta}$ becomes

$$g(\hat{\theta}|\theta) = \frac{1}{(2\pi)^{n/2}|V|^{1/2}} \exp\left[-\tfrac{1}{2}Q(\hat{\theta},\theta)\right], \tag{9.31}$$

where Q is defined as

$$Q(\hat{\theta},\theta) = (\hat{\theta} - \theta)^T V^{-1}(\hat{\theta} - \theta). \tag{9.32}$$

Here V^{-1} is the inverse covariance matrix and the superscript T indicates a transposed (i.e. row) vector. Contours of constant $g(\hat{\theta}|\theta)$ correspond to constant $Q(\hat{\theta},\theta)$. These are ellipses (or for more than two dimensions, hyperellipsoids) in $\hat{\theta}$-space centered about the true parameters θ. Figure 9.7(a) shows a contour of constant $Q(\hat{\theta})$, where θ_{true} represents a particular value of θ.

Also as in the one-dimensional case, one can show that the likelihood function $L(\theta)$ takes on a Gaussian form centered about the ML estimators $\hat{\theta}$,

$$L(\boldsymbol{\theta}) = L_{\max} \exp\left[-\tfrac{1}{2}(\boldsymbol{\theta} - \hat{\boldsymbol{\theta}})^T V^{-1}(\boldsymbol{\theta} - \hat{\boldsymbol{\theta}})\right] = L_{\max} \exp\left[-\tfrac{1}{2} Q(\boldsymbol{\theta}, \hat{\boldsymbol{\theta}})\right]. \quad (9.33)$$

The inverse covariance matrix V^{-1} is the same here as in (9.31); this can be seen from the RCF inequality (6.19) and using the fact that the ML estimators attain the RCF bound in the large sample limit. The quantity Q here is regarded as a function of the parameters $\boldsymbol{\theta}$ which has its maximum at the estimates $\hat{\boldsymbol{\theta}}$. This is shown in Fig. 9.7(b) for $\hat{\boldsymbol{\theta}}$ equal to a particular value $\hat{\boldsymbol{\theta}}_{\mathrm{obs}}$. Because of the symmetry between $\boldsymbol{\theta}$ and $\hat{\boldsymbol{\theta}}$ in the definition (9.32), the quantities Q have the same value in both the p.d.f. (9.31) and in the likelihood function (9.33), i.e. $Q(\hat{\boldsymbol{\theta}}, \boldsymbol{\theta}) = Q(\boldsymbol{\theta}, \hat{\boldsymbol{\theta}})$.

As discussed in Section 7.5, it can be shown that if $\hat{\boldsymbol{\theta}}$ is described by an n-dimensional Gaussian p.d.f. $g(\hat{\boldsymbol{\theta}}, \boldsymbol{\theta})$, then the quantity $Q(\hat{\boldsymbol{\theta}}, \boldsymbol{\theta})$ is distributed according to a χ^2 distribution for n degrees of freedom. The statement that $Q(\hat{\boldsymbol{\theta}}, \boldsymbol{\theta})$ is less than some value Q_γ, i.e. that the estimate is within a certain distance of the true value $\boldsymbol{\theta}$, implies $Q(\boldsymbol{\theta}, \hat{\boldsymbol{\theta}}) < Q_\gamma$, i.e. that the true value $\boldsymbol{\theta}$ is within the same distance of the estimate. The two events therefore have the same probability,

$$P(Q(\boldsymbol{\theta}, \hat{\boldsymbol{\theta}}) \le Q_\gamma) = \int_0^{Q_\gamma} f(z; n)dz, \quad (9.34)$$

where $f(z; n)$ is the χ^2 distribution for n degrees of freedom (equation (2.34)). The value Q_γ is chosen to correspond to a given probability content,

$$\int_0^{Q_\gamma} f(z; n)dz = 1 - \gamma. \quad (9.35)$$

That is,

$$Q_\gamma = F^{-1}(1 - \gamma; n) \quad (9.36)$$

is the quantile of order $1 - \gamma$ of the χ^2 distribution. The region of $\boldsymbol{\theta}$-space defined by $Q(\boldsymbol{\theta}, \hat{\boldsymbol{\theta}}) \le Q_\gamma$ is called a **confidence region** with the confidence level $1 - \gamma$. For a likelihood function of Gaussian form (9.33) it can be constructed by finding the values of $\boldsymbol{\theta}$ at which the log-likelihood function decreases by $Q_\gamma/2$ from its maximum value,

$$\log L(\boldsymbol{\theta}) = \log L_{\max} - \frac{Q_\gamma}{2}. \quad (9.37)$$

As in the single-parameter case, one can still use the prescription given by (9.37) even if the likelihood function is not Gaussian, in which case the probability statement (9.34) is only approximate. For an increasing number of parameters, the approach to the Gaussian limit becomes slower as a function of the sample size, and furthermore it is difficult to quantify when a sample is large enough for (9.34) to apply. If needed, one can determine the probability that a region constructed according to (9.37) includes the true parameter by means of a Monte Carlo calculation.

Quantiles of the χ^2 distribution $Q_\gamma = F^{-1}(1 - \gamma; n)$ for several confidence levels $1 - \gamma$ and $n = 1, 2, 3, 4, 5$ parameters are given in Table 9.4. Values of the confidence level are shown for various values of the quantile Q_γ in Table 9.5.

Table 9.4 The values of the confidence level $1 - \gamma$ for different values of Q_γ and for $n = 1, 2, 3, 4, 5$ fitted parameters.

Q_γ	$1 - \gamma$				
	$n = 1$	$n = 2$	$n = 3$	$n = 4$	$n = 5$
1.0	0.683	0.393	0.199	0.090	0.037
2.0	0.843	0.632	0.428	0.264	0.151
4.0	0.954	0.865	0.739	0.594	0.451
9.0	0.997	0.989	0.971	0.939	0.891

Table 9.5 The values of the quantile Q_γ for different values of the confidence level $1 - \gamma$ for $n = 1, 2, 3, 4, 5$ fitted parameters.

$1 - \gamma$	Q_γ				
	$n = 1$	$n = 2$	$n = 3$	$n = 4$	$n = 5$
0.683	1.00	2.30	3.53	4.72	5.89
0.90	2.71	4.61	6.25	7.78	9.24
0.95	3.84	5.99	7.82	9.49	11.1
0.99	6.63	9.21	11.3	13.3	15.1

For $n = 1$ the expression (9.36) for Q_γ can be shown to imply

$$\sqrt{Q_\gamma} = \Phi^{-1}(1 - \gamma/2), \tag{9.38}$$

where Φ^{-1} is the inverse function of the standard normal distribution. The procedure here thus reduces to that for a single parameter given in Section 9.6, where $N = \sqrt{Q_\gamma}$ is the half-width of the interval in standard deviations (see equations (9.28), (9.29)). The values for $n = 1$ in Tables 9.4 and 9.5 are thus related to those in Tables 9.1 and 9.2 by equation (9.38).

For increasing n, the confidence level for a given Q_γ decreases. For example, in the single-parameter case, $Q_\gamma = 1$ corresponds to $1 - \gamma = 0.683$. For $n = 2$, $Q_\gamma = 1$ gives a confidence level of only 0.393, and in order to obtain $1 - \gamma = 0.683$ one needs $Q_\gamma = 2.30$.

We should emphasize that, as in the single-parameter case, the confidence region $Q(\theta, \hat{\theta}) \leq Q_\gamma$ is a random region in θ-space. The confidence region varies upon repetition of the experiment, since $\hat{\theta}$ is a random variable. The true parameters, on the other hand, are unknown constants.

9.8 Limits near a physical boundary

Often the purpose of an experiment is to search for a new effect, the existence of which would imply that a certain parameter is not equal to zero. For example, one could attempt to measure the mass of the neutrino, which in the standard theory is massless. If the data yield a value of the parameter significantly different from zero, then the new effect has been discovered, and the parameter's value and a confidence interval to reflect its error are given as the result. If, on the other hand, the data result in a fitted value of the parameter that is consistent with zero, then the result of the experiment is reported by giving an upper limit on the parameter. (A similar situation occurs when absence of the new effect corresponds to a parameter being large or infinite; one then places a lower limit. For simplicity we will consider here only upper limits.)

Difficulties arise when an estimator can take on values in the excluded region. This can occur if the estimator $\hat{\theta}$ for a parameter θ is of the form $\hat{\theta} = x - y$, where both x and y are random variables, i.e. they have random measurement errors. The mass squared of a particle, for example, can be estimated by measuring independently its energy E and momentum p, and using $\widehat{m^2} = E^2 - p^2$. Although the mass squared should come out positive, measurement errors in E^2 and p^2 could result in a negative value for $\widehat{m^2}$. Then the question is how to place a limit on m^2, or more generally on a parameter θ when the estimate is in or near an excluded region.

Consider further the example of an estimator $\hat{\theta} = x - y$ where x and y are Gaussian variables with means μ_x, μ_y and variances σ_x^2, σ_y^2. One can show that the difference $\hat{\theta} = x - y$ is also a Gaussian variable with $\theta = \mu_x - \mu_y$ and $\sigma_{\hat{\theta}}^2 = \sigma_x^2 + \sigma_y^2$. (This can be shown using characteristic functions as described in Chapter 10.)

Assume that θ is known a priori to be non-negative (e.g. like the mass squared), and suppose the experiment has resulted in a value $\hat{\theta}_{\mathrm{obs}}$ for the estimator $\hat{\theta}$. According to (9.12), the upper limit θ_{up} at a confidence level $1 - \beta$ is

$$\theta_{\mathrm{up}} = \hat{\theta}_{\mathrm{obs}} + \sigma_{\hat{\theta}}\, \Phi^{-1}(1 - \beta). \tag{9.39}$$

For the commonly used 95% confidence level one obtains from Table 9.2 the quantile $\Phi^{-1}(0.95) = 1.645$.

The interval $(-\infty, \theta_{\mathrm{up}}]$ is constructed to include the true value θ with a probability of 95%, regardless of what θ actually is. Suppose now that the standard deviation is $\sigma_{\hat{\theta}} = 1$, and the result of the experiment is $\hat{\theta}_{\mathrm{obs}} = -2.0$. From equation (9.39) one obtains $\theta_{\mathrm{up}} = -0.355$ at a confidence level of 95%. Not only is $\hat{\theta}_{\mathrm{obs}}$ in the forbidden region (as half of the estimates should be if θ is really zero) but the upper limit is below zero as well. This is not particularly unusual, and in fact is expected to happen in 5% of the experiments if the true value of θ is zero.

As far as the definition of the confidence interval is concerned, nothing fundamental has gone wrong. The interval was designed to cover the true value of θ in a certain fraction of repeated experiments, and we have obviously encountered one of those experiments where θ is not in the interval. But this is not a very satisfying result, since it was already known that θ is greater than zero (and certainly greater than $\theta_{\text{up}} = -0.355$) without having to perform the experiment.

Regardless of the upper limit, it is important to report the actual value of the estimate obtained and its standard deviation, i.e. $\hat{\theta}_{\text{obs}} \pm \sigma_{\hat{\theta}}$, even if the estimate is in the physically excluded region. In this way, the average of many experiments (e.g. as in Section 7.6) will converge to the correct value as long as the estimator is unbiased. In cases where the p.d.f. of $\hat{\theta}$ is significantly non-Gaussian, the entire likelihood function $L(\theta)$ should be given, which can be combined with that of other experiments as discussed in Section 6.12.

Nevertheless, most experimenters want to report some sort of upper limit, and in situations such as the one described above a number of techniques have been proposed (see e.g. [Hig83, Jam91]). There is unfortunately no established convention on how this should be done, and one should therefore state what procedure was used.

As a solution to the difficulties posed by an upper limit in an unphysical region, one might be tempted to simply increase the confidence level until the limit enters the allowed region. In the previous example, if we had taken a confidence level $1 - \beta = 0.99$, then from Table 9.2 one has $\Phi^{-1}(0.99) = 2.326$, giving $\theta_{\text{up}} = 0.326$. This would lead one to quote an upper limit that is smaller than the intrinsic resolution of the experiment ($\sigma_{\hat{\theta}} = 1$) at a very high confidence level of 99%, which is clearly misleading. Worse, of course, would be to adjust the confidence level to give an arbitrarily small limit, e.g. $\Phi^{-1}(0.97725) = 2.00001$, or $\theta_{\text{up}} = 10^{-5}$ at a confidence level of 97.725%!

In order to avoid this type of difficulty, a commonly used technique is to simply shift a negative estimate to zero before applying equation (9.39), i.e.

$$\theta_{\text{up}} = \max(\hat{\theta}_{\text{obs}}, 0) + \sigma_{\hat{\theta}} \, \Phi^{-1}(1 - \beta). \qquad (9.40)$$

In this way the upper limit is always at least the same order of magnitude as the resolution of the experiment. If $\hat{\theta}_{\text{obs}}$ is positive, the limit coincides with that of the classical procedure. This technique has a certain intuitive appeal and is often used, but the interpretation as an interval that will cover the true parameter value with probability $1 - \beta$ no longer applies. The coverage probability is clearly greater than $1 - \beta$, since the shifted upper limit (9.40) is in all cases greater than or equal to the classical one (9.39).

Another alternative is to report an interval based on the Bayesian posterior p.d.f. $p(\theta|\mathbf{x})$. As in Section 6.13, this is obtained from Bayes' theorem,

$$p(\theta|\mathbf{x}) = \frac{L(\mathbf{x}|\theta)\,\pi(\theta)}{\int L(\mathbf{x}|\theta')\,\pi(\theta')d\theta'}, \qquad (9.41)$$

where $\mathbf{x}$ represents the observed data, $L(\mathbf{x}|\theta)$ is the likelihood function and $\pi(\theta)$ is the prior p.d.f. for θ. In Section 6.13, the mode of $p(\theta|\mathbf{x})$ was used as an estimator for θ, and it was shown that this coincides with the ML estimator if the prior density $\pi(\theta)$ is uniform. Here, we can use $p(\theta|\mathbf{x})$ to determine an interval $[a, b]$ such that for given probabilities α and β one has

$$\alpha = \int_{-\infty}^{a} p(\theta|\mathbf{x})\, d\theta$$

$$\beta = \int_{b}^{\infty} p(\theta|\mathbf{x})\, d\theta. \tag{9.42}$$

Choosing $\alpha = \beta$ then gives a central interval, with e.g. $1 - \alpha - \beta = 68.3\%$. Another possibility is to choose α and β such that all values of $p(\theta|\mathbf{x})$ inside the interval $[a, b]$ are higher than any values outside, which implies $p(a|\mathbf{x}) = p(b|\mathbf{x})$. One can show that this gives the shortest possible interval.

One advantage of a Bayesian interval is that prior knowledge, e.g. $\theta \geq 0$, can easily be incorporated by setting the prior p.d.f. $\pi(\theta)$ to zero in the excluded region. Bayes' theorem then gives a posterior probability $p(\theta|\mathbf{x})$ with $p(\theta|\mathbf{x}) = 0$ for $\theta < 0$. The upper limit is thus determined by

$$1 - \beta = \int_{-\infty}^{\theta_{\mathrm{up}}} p(\theta|\mathbf{x})d\theta = \frac{\int_{-\infty}^{\theta_{\mathrm{up}}} L(\mathbf{x}|\theta)\, \pi(\theta)\, d\theta}{\int_{-\infty}^{\infty} L(\mathbf{x}|\theta)\, \pi(\theta)\, d\theta}. \tag{9.43}$$

The difficulties here have already been mentioned in Section 6.13, namely that there is no unique way to specify the prior density $\pi(\theta)$. A common choice is

$$\pi(\theta) = \begin{cases} 0 & \theta < 0 \\ 1 & \theta \geq 0. \end{cases} \tag{9.44}$$

The prescription says in effect: normalize the likelihood function to unit area in the physical region, and then integrate it out to θ_{up} such that the fraction of area covered is $1 - \beta$. Although the method is simple, it has some conceptual drawbacks. For the case where one knows $\theta \geq 0$ (e.g. the neutrino mass) one does not really believe that $0 < \theta < 1$ has the same prior probability as $10^{40} < \theta < 10^{40} + 1$. Furthermore, the upper limit derived from $\pi(\theta) = \text{constant}$ is not invariant with respect to a nonlinear transformation of the parameter.

It has been argued [Jef48] that in cases where $\theta \geq 0$ but with no other prior information, one should use

$$\pi(\theta) = \begin{cases} 0 & \theta \leq 0 \\ \frac{1}{\theta} & \theta > 0. \end{cases} \tag{9.45}$$

This has the advantage that upper limits are invariant with respect to a transformation of the parameter by raising to an arbitrary power. This is equivalent to a uniform (improper) prior of the form (9.44) for $\log \theta$. For this to be usable, however, the likelihood function must go to zero for $\theta \to 0$ and $\theta \to \infty$, or else the integrals in (9.43) diverge. It is thus not applicable in a number of cases of practical interest, including the example discussed in this section. Therefore, despite its conceptual difficulties, the uniform prior density is the most commonly used choice for setting limits on parameters.

Figure 9.8 shows the upper limits at 95% confidence level derived according to the classical, shifted and Bayesian techniques as a function of $\hat{\theta}_{\rm obs} = x - y$ for $\sigma_{\hat{\theta}} = 1$. For the Bayesian limit, a prior density $\pi(\theta) = $ constant was used. The shifted and classical techniques are equal for $\hat{\theta}_{\rm obs} \geq 0$. The Bayesian limit is always positive, and is always greater than the classical limit. As $\hat{\theta}_{\rm obs}$ becomes larger than the experimental resolution $\sigma_{\hat{\theta}}$, the Bayesian and classical limits rapidly approach each other.

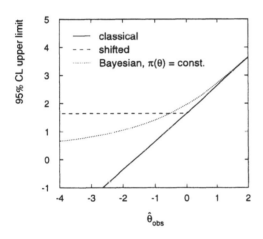

Fig. 9.8 Upper limits at 95% confidence level for the example of Section 9.8 using the classical, shifted and Bayesian techniques. The shifted and classical techniques are equal for $\hat{\theta}_{\rm obs} \geq 0$.

9.9 Upper limit on the mean of Poisson variable with background

As a final example, recall Section 9.4 where an upper limit was placed on the mean ν of a Poisson variable n. Often one is faced with a somewhat more complicated situation where the observed value of n is the sum of the desired signal events $n_{\rm s}$ as well as background events $n_{\rm b}$,

$$n = n_{\rm s} + n_{\rm b}, \tag{9.46}$$

where both $n_{\rm s}$ and $n_{\rm b}$ can be regarded as Poisson variables with means $\nu_{\rm s}$ and $\nu_{\rm b}$, respectively. Suppose for the moment that the mean for the background $\nu_{\rm b}$ is known without any uncertainty. For $\nu_{\rm s}$ one only knows a priori that $\nu_{\rm s} \geq 0$. The goal is to construct an upper limit for the signal parameter $\nu_{\rm s}$ given a measured value of n.

Since n is the sum of two Poisson variables, one can show that it is itself a Poisson variable, with the probability function

$$f(n; \nu_s, \nu_b) = \frac{(\nu_s + \nu_b)^n}{n!} e^{-(\nu_s + \nu_b)}. \tag{9.47}$$

The ML estimator for ν_s is

$$\hat{\nu}_s = n - \nu_b, \tag{9.48}$$

which has zero bias since $E[n] = \nu_s + \nu_b$. Equations (9.15), which are used to determine the confidence interval, become

$$
\begin{aligned}
\alpha &= P(\hat{\nu}_s \geq \hat{\nu}_s^{obs}; \nu_s^{lo}) = \sum_{n \geq n_{obs}} \frac{(\nu_s^{lo} + \nu_b)^n \, e^{-(\nu_s^{lo} + \nu_b)}}{n!}, \\
\beta &= P(\hat{\nu}_s \leq \hat{\nu}_s^{obs}; \nu_s^{up}) = \sum_{n \leq n_{obs}} \frac{(\nu_s^{up} + \nu_b)^n \, e^{-(\nu_s^{up} + \nu_b)}}{n!}.
\end{aligned}
\tag{9.49}
$$

These can be solved numerically for the lower and upper limits ν_s^{lo} and ν_s^{up}. Comparing with the case $\nu_b = 0$, one sees that the limits from (9.49) are related to what would be obtained without background by

$$
\begin{aligned}
\nu_s^{lo} &= \nu_s^{lo}(\text{no background}) - \nu_b, \\
\nu_s^{up} &= \nu_s^{up}(\text{no background}) - \nu_b.
\end{aligned}
\tag{9.50}
$$

The difficulties here are similar to those encountered in the previous example. The problem occurs when the total number of events observed n_{obs} is not large compared to the expected number of background events ν_b. Values of ν_s^{up} for $1 - \beta = 0.95$ are shown in Fig. 9.9(a) as a function of the expected number of background events ν_b. For small enough n_{obs} and a high enough background level ν_b, a non-negative solution for ν_s^{up} does not exist. This situation can occur, of course, because of fluctuations in n_s and n_b.

Because of these difficulties, the classical limit is not recommended in this case. As previously mentioned, one should always report $\hat{\nu}_s$ and an estimate of its variance even if $\hat{\nu}_s$ comes out negative. In this way the average of many experiments will converge to the correct value. If, in addition, one wishes to report an upper limit on ν_s, the Bayesian method can be used with, for example, a uniform prior density [Hel83]. The likelihood function is given by the probability (9.47), now regarded as a function of ν_s,

$$L(n_{obs}|\nu_s) = \frac{(\nu_s + \nu_b)^{n_{obs}}}{n_{obs}!} e^{-(\nu_s + \nu_b)}. \tag{9.51}$$

The posterior probability density for ν_s is obtained as usual from Bayes' theorem,

$$p(\nu_s|n_{obs}) = \frac{L(n_{obs}|\nu_s) \, \pi(\nu_s)}{\int_{-\infty}^{\infty} L(n_{obs}|\nu_s') \, \pi(\nu_s') \, d\nu_s'}. \tag{9.52}$$

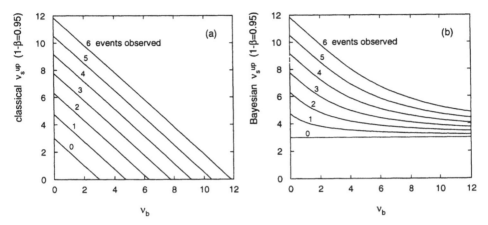

Fig. 9.9 Upper limits ν_s^{up} at a confidence level of $1 - \beta = 0.95$ for different numbers of events observed n_{obs} and as a function of the expected number of background events ν_b. (a) The classical limit. (b) The Bayesian limit based on a uniform prior density for ν_s.

Taking $\pi(\nu_s)$ to be constant for $\nu_s \geq 0$ and zero for $\nu_s < 0$, the upper limit ν_s^{up} at a confidence level of $1 - \beta$ is given by

$$1 - \beta = \frac{\int_0^{\nu_s^{up}} L(n_{obs}|\nu_s)\, d\nu_s}{\int_0^{\infty} L(n_{obs}|\nu_s)\, d\nu_s}$$

$$= \frac{\int_0^{\nu_s^{up}} (\nu_s + \nu_b)^{n_{obs}} e^{-(\nu_s + \nu_b)}\, d\nu_s}{\int_0^{\infty} (\nu_s + \nu_b)^{n_{obs}} e^{-(\nu_s + \nu_b)}\, d\nu_s}. \tag{9.53}$$

The integrals can be related to incomplete gamma functions (see e.g. [Arf95]), or since n_{obs} is a positive integer, they can be solved by making the substitution $x = \nu_s + \nu_b$ and integrating by parts n_{obs} times. Equation (9.53) then becomes

$$\beta = \frac{e^{-(\nu_s^{up} + \nu_b)} \sum_{n=0}^{n_{obs}} \frac{(\nu_s^{up} + \nu_b)^n}{n!}}{e^{-\nu_b} \sum_{n=0}^{n_{obs}} \frac{\nu_b^n}{n!}}. \tag{9.54}$$

This can be solved numerically for the upper limit ν_s^{up}. The upper limit as a function of ν_b is shown in Fig. 9.9(b) for various values of n_{obs}. For the case without background, setting $\nu_b = 0$ gives

$$\beta = e^{-\nu_s^{up}} \sum_{n=0}^{n_{obs}} \frac{(\nu_s^{up})^n}{n!}, \tag{9.55}$$

which is identical to the equation for the classical upper limit (9.16). This can be seen by comparing Figs 9.9(a) and (b). The Bayesian limit is always greater than or equal to the corresponding classical one, with the two agreeing only for $\nu_b = 0$.

The agreement for the case without background must be considered accidental, however, since the Bayesian limit depends on the particular choice of a constant prior density $\pi(\nu_s)$. Nevertheless, the coincidence spares one the trouble of having to defend either the classical or Bayesian viewpoint, which may account for the general acceptance of the uniform prior density in this case.

Often the result of an experiment is not simply the number n of observed events, but includes in addition measured values $x_1, \ldots, x_n$ of some property of the events. Suppose the probability density for x is

$$f(x; \nu_s, \nu_b) = \frac{\nu_s f_s(x) + \nu_b f_b(x)}{\nu_s + \nu_b}, \tag{9.56}$$

where the components $f_s(x)$ for signal and $f_b(x)$ for background events are both assumed to be known. If these p.d.f.s have different shapes, then the values of x contain additional information on whether the observed events were signal or background. This information can be incorporated into the limit ν_s by using the extended likelihood function,

$$
\begin{aligned}
L(\nu_s) &= \frac{(\nu_s + \nu_b)^n}{n!} e^{-(\nu_s + \nu_b)} \prod_{i=1}^{n} \frac{\nu_s f_s(x_i) + \nu_b f_b(x_i)}{\nu_s + \nu_b} \\
&= \frac{e^{-(\nu_s + \nu_b)}}{n!} \prod_{i=1}^{n} [\nu_s f_s(x_i) + \nu_b f_b(x_i)], \tag{9.57}
\end{aligned}
$$

as defined in Section 6.9, or by using the corresponding formula for binned data as discussed in Section 6.10.

In the classical case, one uses the likelihood function to find the estimator $\hat{\nu}_s$. In order to find the classical upper limit, however, one requires the p.d.f. of $\hat{\nu}_s$. This is no longer as simple to find as before, where only the number of events was counted, and must in general be determined numerically. For example, one can perform Monte Carlo experiments using a given value of ν_s (and the known value ν_b) to generate numbers n_s and n_b from a Poisson distribution, and corresponding x values according to $f_s(x; \nu_s)$ and $f_b(x; \nu_b)$. By adjusting ν_s, one can find that value for which there is a probability β to obtain $\hat{\nu}_s \leq \hat{\nu}_s^{obs}$. Here one must still deal with the problem that the limit can turn out negative.

In the Bayesian approach, $L(\nu_s)$ is used directly in Bayes' theorem as before. Solving equation (9.53) for ν_s^{up} must in general be done numerically. This has the advantage of not requiring the sampling p.d.f. for the estimator $\hat{\nu}_s$, in addition to the previously mentioned advantage of automatically incorporating the prior knowledge $\nu_s \geq 0$ into the limit.

Further discussion of the issue of Bayesian versus classical limits can be found in [Hig83, Jam91, Cou95]. A technique for incorporating systematic uncertainties in the limit is given in [Cou92].

10
Characteristic functions and related examples

10.1 Definition and properties of the characteristic function

The **characteristic function** $\phi_x(k)$ for a random variable x with p.d.f. $f(x)$ is defined as the expectation value of e^{ikx},

$$\phi_x(k) = E[e^{ikx}] = \int_{-\infty}^{\infty} e^{ikx} f(x)dx. \tag{10.1}$$

This is essentially the Fourier transform of the probability density function. It is useful in proving a number of important theorems, in particular those involving sums of random variables. One can show that there is a one-to-one correspondence between the p.d.f. and the characteristic function, so that knowledge of one is equivalent to knowledge of the other. Some characteristic functions of important p.d.f.s are given in Table 10.1.

Suppose one has n independent random variables $x_1, \ldots, x_n$, with p.d.f.s $f_1(x_1), \ldots, f_n(x_n)$, and corresponding characteristic functions $\phi_1(k), \ldots, \phi_n(k)$, and consider the sum $z = \sum_i x_i$. The characteristic function $\phi_z(k)$ for z is related to those of the x_i by

$$
\begin{aligned}
\phi_z(k) &= \int \ldots \int \exp\left(ik\sum_{i=1}^{n} x_i\right) f_1(x_1) \ldots f_n(x_n)dx_1 \ldots dx_n \\
&= \int e^{ikx_1} f_1(x_1)dx_1 \ldots \int e^{ikx_n} f_n(x_n)dx_n \\
&= \phi_1(k) \ldots \phi_n(k).
\end{aligned}
\tag{10.2}
$$

That is, the characteristic function for a sum of independent random variables is given by the product of the individual characteristic functions.

The p.d.f. $f(z)$ is obtained from the inverse Fourier transform,

$$f(z) = \frac{1}{2\pi} \int_{-\infty}^{\infty} \phi_z(k) e^{-ikz}\, dk. \tag{10.3}$$

Table 10.1 Characteristic functions for some commonly used probability functions.

Distribution	p.d.f.	$\phi(k)$
Binomial	$f(n;N,p) = \frac{N!}{n!(N-n)!} p^n (1-p)^{N-n}$	$[p(e^{ik}-1)+1]^N$
Poisson	$f(n;\nu) = \frac{\nu^n}{n!} e^{-\nu}$	$\exp[\nu(e^{ik}-1)]$
Uniform	$f(x;\alpha,\beta) = \begin{cases} \frac{1}{\beta-\alpha} & \alpha \le x \le \beta \\ 0 & \text{otherwise} \end{cases}$	$\frac{e^{i\beta k}-e^{i\alpha k}}{(\beta-\alpha)ik}$
Exponential	$f(x;\xi) = \frac{1}{\xi}e^{-x/\xi}$	$\frac{1}{1-ik\xi}$
Gaussian	$f(x;\mu,\sigma^2) = \frac{1}{\sqrt{2\pi\sigma^2}} \exp\left(\frac{-(x-\mu)^2}{2\sigma^2}\right)$	$\exp(i\mu k - \frac{1}{2}\sigma^2 k^2)$
Chi-square	$f(z;n) = \frac{1}{2^{n/2}\Gamma(n/2)} z^{n/2-1} e^{-z/2}$	$(1-2ik)^{-n/2}$
Cauchy	$f(x) = \frac{1}{\pi}\frac{1}{1+x^2}$	$e^{-|k|}$

Even if one is unable to invert the transform to find $f(z)$, one can easily determine its moments. Differentiating the characteristic function m times gives

$$
\begin{aligned}
\left.\frac{d^m}{dk^m}\phi_z(k)\right|_{k=0} &= \left.\frac{d^m}{dk^m}\int e^{ikz} f(z)dz\right|_{k=0} \\
&= i^m \int z^m f(z)dz \\
&= i^m \mu'_m \quad\quad\quad (10.4)
\end{aligned}
$$

where $\mu'_m = E[z^m]$ is the mth algebraic moment of z.

10.2   Applications of the characteristic function

In this section we will demonstrate the use of the characteristic function by deriving a number of results that have been stated without proof in earlier chapters.

As a first example, we can use equation (10.4) to determine the means, variances and higher moments of the various distributions introduced in Chapter 2. The mean and variance of the Gaussian distribution, for example, are

$$
E[x] \;=\; -i\frac{d}{dk}[\exp(i\mu k - \tfrac{1}{2}\sigma^2 k^2)]\Big|_{k=0} = \mu,
$$

$$
V[x] \;=\; E[x^2] - (E[x])^2
$$

$$
\;=\; -\frac{d^2}{dk^2}[\exp(i\mu k - \tfrac{1}{2}\sigma^2 k^2)]\Big|_{k=0} - \mu^2 = \sigma^2. \tag{10.5}
$$

In a similar way one can find the moments for the other distributions listed in Table 10.1, with the exception of the Cauchy distribution. Here the characteristic function $\phi(k) = e^{-|k|}$ is not differentiable at $k = 0$, and as noted in Section 2.8, the mean and higher moments do not exist.

By using characteristic functions it is easy to investigate how distributions behave for certain limiting cases of their parameters. For the binomial distribution, for example, the characteristic function is

$$
\phi(k) \;=\; [p(e^{ik} - 1) + 1]^N. \tag{10.6}
$$

Taking the limit $p \to 0$, $N \to \infty$ with $\nu = pN$ constant gives

$$
\phi(k) \;=\; \left(\frac{\nu}{N}(e^{ik} - 1) + 1\right)^N \to \exp[\nu(e^{ik} - 1)], \tag{10.7}
$$

which is the characteristic function of the Poisson distribution.

In a similar way, one can show that a Poisson variable n with mean ν becomes a Gaussian variable in the limit $\nu \to \infty$. Although the Poisson variable is discrete, for large n it can be treated as a continuous variable x as long as it is integrated over an interval that is large compared to unity. Recall that the variance of a Poisson variable is equal to its mean ν. For convenience we can transform from n to the variable

$$
x = \frac{n - \nu}{\sqrt{\nu}}, \tag{10.8}
$$

which has a mean of zero and a variance of unity. The characteristic function of x is

$$
\phi_x(k) \;=\; E[e^{ikx}] = E\left[e^{ikn/\sqrt{\nu}}\, e^{-ik\sqrt{\nu}}\right]
$$

$$
\;=\; \phi_n\left(\frac{k}{\sqrt{\nu}}\right) e^{-ik\sqrt{\nu}}, \tag{10.9}
$$

where ϕ_n is the characteristic function of the Poisson distribution. Substituting this from Table 10.1, expanding the exponential and taking the limit $\nu \to \infty$, equation (10.9) becomes

$$\phi_x(k) = \exp\left[\nu\left(e^{ik/\sqrt{\nu}} - 1\right) - ik\sqrt{\nu}\right] \to \exp\left(-\tfrac{1}{2}k^2\right). \tag{10.10}$$

This, however, is the characteristic function for a Gaussian with a mean of zero and unit variance. Transforming back to the original Poisson variable n, one finds that for large ν, n follows a Gaussian distribution with mean and variance both equal to ν.

The addition theorem (10.2) allows us to prove a number of important results. For example, consider the sum z of two Gaussian random variables x and y with means μ_x, μ_y and variances σ_x^2, σ_y^2. According to (10.2) the characteristic function for z is related to those of x and y by

$$
\begin{aligned}
\phi_z(k) &= \phi_x(k)\,\phi_y(k) \\
&= \exp(i\mu_x k - \tfrac{1}{2}\sigma_x^2 k^2) \cdot \exp(i\mu_y k - \tfrac{1}{2}\sigma_y^2 k^2) \\
&= \exp[i(\mu_x + \mu_y)k - \tfrac{1}{2}(\sigma_x^2 + \sigma_y^2)].
\end{aligned}
\tag{10.11}
$$

This shows that z is itself a Gaussian random variable with mean $\mu_z = \mu_x + \mu_y$ and variance $\sigma_z^2 = \sigma_x^2 + \sigma_y^2$. The corresponding property for the difference of two Gaussian variables was used in the example of Section 9.8. In the same way one can show that the sum of Poisson variables with means ν_i is itself a Poisson variable with mean $\sum_i \nu_i$.

Also using (10.2) one can show that for n independent Gaussian random variables x_i with means μ_i and variances σ_i^2, the sum of squares

$$z = \sum_{i=1}^{n} \frac{(x_i - \mu_i)^2}{\sigma_i^2} \tag{10.12}$$

follows a χ^2 distribution for n degrees of freedom. To see this, consider first the case $n = 1$. By transformation of variables, one can show that

$$y = \frac{x_i - \mu_i}{\sigma_i} \tag{10.13}$$

follows the standard Gaussian p.d.f. $\varphi(y) = (1/\sqrt{2\pi})e^{-y^2/2}$ for all i, and that $z = y^2$ therefore follows

$$f(z; n = 1) = 2\varphi(y)\left|\frac{dy}{dz}\right| = \frac{1}{\sqrt{2\pi z}}e^{-z/2}, \tag{10.14}$$

where the factor of two is necessary to account for both positive and negative values of y. Comparison with (2.34) shows that this is the χ^2 distribution for $n = 1$. From Table 10.1, the characteristic function for z is

$$\phi(k) = (1 - 2ik)^{-1/2}. \tag{10.15}$$

For a sum of n terms, i.e. $z = \sum_{i=1}^{n} y^2$, the characteristic function is the product of n identical factors like (10.15), which gives directly the characteristic function of the χ^2 distribution for n degrees of freedom.

A proof of the central limit theorem based on similar arguments is sufficiently important to merit a separate discussion; this is given in the following section.

10.3 The central limit theorem

Suppose we have n independent random variables x_j with means μ_j and variances σ_j^2, not necessarily equal. The central limit theorem states that in the limit of large n, the sum $\sum_j x_j$ becomes a Gaussian random variable with mean $\sum_j \mu_j$ and variance $\sum_j \sigma_j^2$. This holds under fairly general conditions regardless of distributions of the individual x_j.

For convenience, we can subtract off the means μ_j so that the variables all have mean values of zero. In addition, we can regard n for the moment as fixed, and define

$$y_j = \frac{x_j - \mu_j}{\sqrt{n}}, \tag{10.16}$$

so that $E[y_j] = 0$ and $E[y_j^2] = \sigma_j^2/n$. Consider the case where all of the variances are equal, $\sigma_j^2 = \sigma^2$. The characteristic function $\phi_j(k)$ for y_j can be expanded in a Taylor series as

$$
\begin{aligned}
\phi_j(k) &= \sum_{m=0}^{\infty} \frac{d^m \phi}{dk^m}\bigg|_{k=0} \frac{k^m}{m!} \\
&= \sum_{m=0}^{\infty} \frac{(ik)^m}{m!} E[y^m] \\
&= 1 - \frac{k^2}{2n}\sigma^2 - \frac{ik^3}{3!} \frac{E[(x_j - \mu_j)^3]}{n^{3/2}} + \cdots, \tag{10.17}
\end{aligned}
$$

where we have used equation (10.4) to relate the derivatives $d^m \phi_j/dk^m$ to the moments $E[y_j^m]$. By using equation (10.2), the characteristic function $\phi_z(k)$ for the sum $z = \sum_j y_j$ is

$$\phi_z(k) = \prod_{j=1}^{n} \phi_j(k) = \prod_{j=1}^{n} \left(1 - \frac{k^2}{2n}\sigma^2 - \frac{ik^3}{3!} \frac{E[(x_j - \mu_j)^3]}{n^{3/2}} + \cdots \right). \tag{10.18}$$

If the terms with the third and higher moments can be neglected in the limit of large n, one obtains

$$\phi_z(k) \approx \left(1 - \frac{k^2}{2n}\sigma^2\right)^n \to \exp\left(-\tfrac{1}{2}\sigma^2 k^2\right). \tag{10.19}$$

This is the characteristic function of a Gaussian with mean zero and variance σ^2. By transforming back to the variable $\sum_j x_j$ one obtains a Gaussian with mean $\sum_j \mu_j$ and variance $n\sigma^2$. The theorem holds as well for the case where the σ_j are different, under somewhat more restrictive conditions, with the variance of the sum then being $\sum_j \sigma_j^2$.

Rather than specify the conditions under which the central limit theorem holds exactly in the limit $n \to \infty$ (see e.g. [Gri92]), it is more important in a practical data analysis to know to what extent the Gaussian approximation is valid for finite n. This is difficult to quantify, but one can say roughly that it holds as long as the sum is built up of a large number of small contributions. Discrepancies arise if, for example, the distributions of the individual terms have long tails, so that occasional large values make up a large part of the sum. Such contributions lead to 'non-Gaussian' tails in the sum, which can significantly alter the probability to find values with large departures from the mean.

A common implicit use of the central limit theorem is the assumption that the measured value of a quantity is a Gaussian distributed variable centered about the true value. This assumption is often used when constructing a confidence interval, cf. Chapter 9. Such intervals can be significantly underestimated if non-Gaussian tails are present. In particular, the relationship between the confidence level and the size of the interval will differ from the Gaussian prescription, equation (9.12), e.g. 68.3% for a '1 σ' interval, 95.4% for 2 σ, etc. A better understanding of the non-Gaussian tails can sometimes be obtained from a detailed Monte Carlo simulation of the individual variables making up the sum.

An example where the central limit theorem breaks down is the total number of electron–ion pairs created when a charged particle traverses a layer of matter. The number of pairs in a layer of a given thickness can be described by the Landau distribution, seen in Section 2.9. This was shown in Fig. 2.9 for a 4 mm layer of argon gas. The total amount of ionization in traversing, say, a 1 meter gas volume is then the sum of 25 such variables. In general, if one considers the total volume as being subdivided into a large number of very thin layers, then the total ionization is the sum of a large number of individual contributions, and one would expect the central limit theorem to apply. But the Landau distribution has a long tail extending to large values, so that relatively rare highly ionizing collisions can make up a significant fraction of the total ionization. The Gaussian approximation is not in general valid in practical systems (cf. [All80]).

Another example is the angle by which a charged particle is deflected upon traversing a layer of matter. The total angle can be regarded as the sum of a small number of deflections caused by collisions with nuclei in the substance being traversed (multiple Coulomb scattering). Since there are many such collisions, one expects a Gaussian distribution for the total angle. The distribution for the individual collisions, however, has a long tail extending to large angles. For a finite thickness, rare collisions leading to large angles can make up a significant fraction of the total, leading to a non-Gaussian distribution for the final angle.

10.4 Use of the characteristic function to find the p.d.f. of an estimator

Consider n independent observations of a random variable x from an exponential distribution $f(x; \xi) = (1/\xi) \exp(-x/\xi)$. In Section 6.2 it was seen that the maximum likelihood estimator $\hat{\xi}$ for ξ was the sample mean of the observed x_i:

$$\hat{\xi} = \frac{1}{n} \sum_{i=1}^{n} x_i. \tag{10.20}$$

If the experiment were repeated many times one would obtain values of $\hat{\xi}$ distributed according to a p.d.f. $g(\hat{\xi}; n, \xi)$ which depends on the number of observations per experiment n and the true value of the parameter ξ.

Suppose we want to find $g(\hat{\xi}; n, \xi)$. The characteristic function for x is

$$\phi_x(k) = \int e^{ikx} f(x) dx$$

$$= \int_0^\infty e^{ikx} \frac{1}{\xi} e^{-x/\xi} dx$$

$$= \frac{1}{1 - ik\xi}. \tag{10.21}$$

Applying equation (10.2) for the sum $z = \sum_{i=1}^{n} x_i = n\hat{\xi}$ gives

$$\phi_z(k) = \frac{1}{(1 - ik\xi)^n}. \tag{10.22}$$

The p.d.f. $g_z(z)$ for z is found by computing the inverse Fourier transform of $\phi_z(k)$,

$$g_z(z) = \frac{1}{2\pi} \int_{-\infty}^{\infty} \frac{e^{-ikz}}{(1 - ik\xi)^n} dk. \tag{10.23}$$

The integrand has a pole of order n at $-i/\xi$ in the complex k plane. Closing the contour in the lower half plane and using the residue theorem gives

$$g_z(z) = \frac{1}{(n-1)!} \frac{z^{n-1}}{\xi^n} e^{-z/\xi}. \tag{10.24}$$

Transforming to find the p.d.f. for the estimator $\hat{\xi} = z/n$ gives

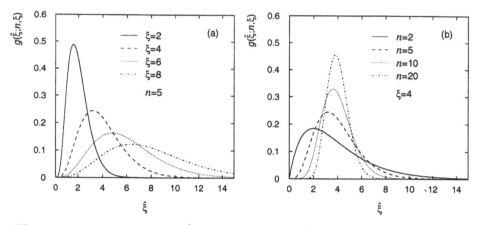

Fig. 10.1 The sampling p.d.f. $g(\hat{\xi}; n, \xi)$ for the estimator $\hat{\xi}$ for various values of n and ξ. (a) $n = 5$ measurements and various values of the true parameter ξ. (b) $\xi = 4$ and various numbers of measurements n.

$$g(\hat{\xi}; n, \xi) = g_z(z) \left| dz/d\hat{\xi} \right|$$

$$= ng_z(n\hat{\xi})$$

$$= \frac{n^n}{(n-1)!} \frac{\hat{\xi}^{n-1}}{\xi^n} e^{-n\hat{\xi}/\xi}, \qquad (10.25)$$

which is a special case of the gamma distribution (see e.g. [Ead71] Chapter 4). Figure 10.1 shows the distribution $g(\hat{\xi}; n, \xi)$ for several values of the parameters. For $n = 5$ measurements one sees that the p.d.f. is roughly centered about the true value ξ, but has a long tail extending to higher values of $\hat{\xi}$. In Fig. 10.1(b) one sees that the p.d.f. becomes approximately Gaussian as the number of measurements n increases, as required by the central limit theorem.

We can now take advantage of the fact that we have the p.d.f. of an estimator to work out two additional examples. In Section 10.4.1 the p.d.f. (10.25) is used to compute expectation values, and in Section 10.4.2 it is used to construct a confidence interval.

10.4.1   Expectation value for mean lifetime and decay constant

Changing now to the conventional notation for particle lifetimes, equation (10.25) gives the p.d.f. of $\hat{\tau} = (1/n) \sum_{i=1}^{n} t_i$ used to estimate the mean lifetime τ of a particle given n decay-time measurements $t_1, \dots, t_n$. Recall that the expectation value of $\hat{\tau}$ was computed in Section 6.2 by using the formula

$$E[\hat{\tau}(t_1,\ldots,t_n)] = \int_0^\infty \cdots \int_0^\infty \left(\frac{1}{n}\sum_{i=1}^n t_i\right) \frac{1}{\tau} e^{-t_1/\tau} \cdots \frac{1}{\tau} e^{-t_n/\tau} dt_1 \ldots dt_n = \tau.$$

(10.26)

This result could have also been obtained directly from the p.d.f. of $\hat{\tau}$ (see equation (10.25)),

$$
\begin{aligned}
E[\hat{\tau}] &= \int_0^\infty \hat{\tau}\, g(\hat{\tau}; n, \tau)\, d\hat{\tau} \\
&= \int_0^\infty \hat{\tau}\, \frac{n^n}{(n-1)!} \frac{\hat{\tau}^{n-1}}{\tau^n} e^{-n\hat{\tau}/\tau}\, d\hat{\tau} \\
&= \tau.
\end{aligned}
$$

(10.27)

It was also shown in Section 6.2 that the maximum likelihood estimator for a function of a parameter is given by the same function of the ML estimator for the original parameter. For example, the ML estimator for the decay constant $\lambda = 1/\tau$ is $\hat{\lambda} = 1/\hat{\tau}$. From $g(\hat{\tau}; n, \tau)$ one can compute the p.d.f. $h(\hat{\lambda})$,

$$
\begin{aligned}
h(\hat{\lambda}; n, \lambda) &= g(\hat{\tau}; n, \tau) \left| d\hat{\tau}/d\hat{\lambda} \right| \\
&= \frac{n^n}{(n-1)!} \frac{\lambda^n}{\hat{\lambda}^{n+1}} e^{-n\lambda/\hat{\lambda}}.
\end{aligned}
$$

(10.28)

The expectation value of $\hat{\lambda}$ is

$$
\begin{aligned}
E[\hat{\lambda}] &= \int_0^\infty \hat{\lambda}\, h(\hat{\lambda}; n, \lambda)\, d\hat{\lambda} \\
&= \int_0^\infty \frac{n^n}{(n-1)!} \frac{\lambda^n}{\hat{\lambda}^n} e^{-n\lambda/\hat{\lambda}}\, d\hat{\lambda} \\
&= \frac{n}{n-1} \lambda.
\end{aligned}
$$

(10.29)

One sees that even though the ML estimator $\hat{\tau} = (1/n)\sum_{i=1}^n t_i$ is an unbiased estimator for τ, the estimator $\hat{\lambda} = 1/\hat{\tau}$ is not an unbiased estimator for $\lambda = 1/\tau$. The bias, however, goes to zero in the limit that n goes to infinity.

10.4.2 Confidence interval for the mean of an exponential random variable

The p.d.f. $g(\hat{\xi}; n, \xi)$ from equation (10.25) can be used to determine a confidence interval according to the procedure given in Section 9.2. Suppose n observations of the exponential random variable x have been used to evaluate the estimator $\hat{\xi}$

for the parameter ξ, and the value obtained is $\hat{\xi}_{\text{obs}}$. The goal is to determine an interval $[a, b]$ given the data $x_1, \ldots, x_n$ such that the probabilities $P[a < \xi] = \alpha$ and $P[\xi < b] = \beta$ hold for fixed α and β regardless of the true value ξ.

The confidence interval is found by solving equations (9.9) for a and b,

$$\alpha = \int_{\hat{\xi}_{\text{obs}}}^{\infty} g(\hat{\xi}; a) \, d\hat{\xi},$$

$$\beta = \int_{-\infty}^{\hat{\xi}_{\text{obs}}} g(\hat{\xi}; b) \, d\hat{\xi}. \tag{10.30}$$

Figure 10.2 shows the 68.3% confidence intervals for various values of n assuming a measured value $\hat{\xi}_{\text{obs}} = 1$. Also shown are the intervals one would obtain from the measured value plus or minus the estimated standard deviation. As n becomes larger the p.d.f. $g(\hat{\xi}; n, \xi)$ becomes Gaussian (as it must by the central limit theorem) and the 68.3% central confidence interval approaches $[\hat{\xi}_{\text{obs}} - \hat{\sigma}_{\hat{\xi}}, \hat{\xi}_{\text{obs}} + \hat{\sigma}_{\hat{\xi}}]$.

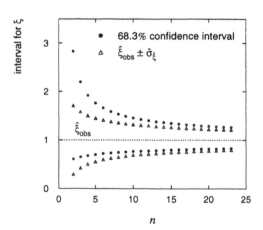

Fig. 10.2 Classical confidence intervals for the parameter of the exponential distribution ξ (between solid points) and the interval $[\hat{\xi}_{\text{obs}} - \hat{\sigma}_{\hat{\xi}}, \hat{\xi}_{\text{obs}} + \hat{\sigma}_{\hat{\xi}}]$ (between open triangles) for different values of the number of measurements n, assuming an observed value $\hat{\xi}_{\text{obs}} = 1$.

11
Unfolding

Up to now we have considered random variables such as particle energies, decay times, etc., usually with the assumption that their values can be measured without error. The present chapter concerns the distortions to distributions which occur when the values of these variables are subject to additional random fluctuations due to the limited resolution of the measuring device. The procedure of correcting for these distortions is known as **unfolding**. The same mathematics can be found under the general heading of **inverse problems**, and is also called **deconvolution** or **unsmearing**. Although the presentation here is mainly in the context of particle physics, the concepts have been developed and applied in fields such as optical image reconstruction, radio astronomy, crystallography and medical imaging.

The approach here, essentially that of classical statistics, follows in many ways that of [Any91, Any92, Bel85, Zhi83, Zhi88]. Some of the methods have a Bayesian motivation as well, however, cf. [Siv96, Ski85, Ski86, Jay86].

In Section 11.1 the unfolding problem is formulated and the notation defined. Unfolding by inversion of the response matrix is discussed in Section 11.2. This technique is rarely used in practice, but is a starting point for better solutions. A simple method based on correction factors is shown in Section 11.3. The main topic of this chapter, regularized unfolding, is described in Sections 11.4 through 11.7. This includes the strategy used to find the solution, a survey of several regularization functions, and methods for estimating the variance and bias of the solution. These points are illustrated by means of examples in Section 11.8, and information on numerical implementation of the methods is given in Section 11.9.

It should be emphasized that in many problems it is not necessary to unfold the measured distribution, in particular if the goal is to compare the result with the prediction of an existing theory. In that case one can simply modify the prediction to include the distortions of the detector, and this can be directly compared with the measurement. This procedure is considerably simpler than unfolding the measurement and comparing it with the original (unmodified) theory.

Without unfolding, however, the measurement cannot be compared with the results of other experiments, in which the effects of resolution will in general be different. It can also happen that a new theory is developed many years after a measurement has been carried out, and the information needed to modify the theory for the effects of resolution, i.e. the response function or matrix (see below), may no longer be available. If a particularly important measured distribution is to retain its value, then both the measurement and the response matrix should be preserved. Unfortunately, this is often impractical, and it is rarely done.

By unfolding the distribution one provides a result which can directly be compared with those of other experiments as well as with theoretical predictions. Other reasons for unfolding exist in applications such as image reconstruction, where certain features may not be recognizable in the uncorrected distribution. In this chapter we will assume that these arguments have been considered and that the decision has been made to unfold.

11.1   Formulation of the unfolding problem

Consider a random variable x whose p.d.f. we would like to determine. In this chapter we will allow for limited resolution in the measurement of x, as well as detection efficiency less than 100% and the presence of background processes. As an example, we could consider the distribution of electron energies resulting from the beta decay of radioactive nuclei, i.e. the variable x refers to the energy of the emitted electron.

By 'limited resolution' we mean that because of measurement errors, the measured values of x may differ in a random way from the values that were actually created. For example, a particular beta decay may result in an electron with a certain energy, but because of the resolution of the measuring device, the recorded value will in general be somewhat different. Each observed event is thus characterized by two quantities: a true value y (which we do not know) and an observed value x.

In general one must also allow for the occurrence of a true value that does not result in any measured value at all. For the example of beta decay, it could be that an emitted electron escapes completely undetected, since the detector may not cover the entire solid angle surrounding the radioactive source, or electron energies below a certain minimum threshold may not produce a sufficiently large signal to be detected. The probability that an event leads to some measured value is called the detection **efficiency**[1] $\varepsilon(y)$, which in general depends on the true value of the event, y.

Suppose the true values are distributed according to the p.d.f. $f_{\text{true}}(y)$. In order to construct a usable estimator for $f_{\text{true}}(y)$, it is necessary to represent it by means of some finite set of parameters. If no functional form for $f_{\text{true}}(y)$ is known a priori, then it can still be represented as a normalized histogram with M bins. The probability to find y in bin j is simply the integral over the bin,

---

[1]If the reason that the event went undetected is related to the geometry, e.g. limited solid angle of the detector, then the efficiency is often called acceptance. The term efficiency is sometimes used to refer to the conditional probability that an event is detected given that it is contained in the sensitive region of the detector. Here we will use efficiency in the more general sense, meaning the overall probability for an event to be detected.

$$p_j = \int_{\text{bin } j} f_{\text{true}}(y) \, dy. \tag{11.1}$$

Suppose we perform an experiment in which a certain total number of events m_{tot} occur; this will differ in general from the number observed. The number m_{tot} could be treated as fixed or as a random variable. In either case, we will call the expectation value of the total number of events $\mu_{\text{tot}} = E[m_{\text{tot}}]$, so that the expected number of events in bin j is

$$\mu_j = \mu_{\text{tot}} \, p_j. \tag{11.2}$$

We will refer to the vector $\boldsymbol{\mu} = (\mu_1, \ldots, \mu_M)$ as the 'true histogram'. Note that these are not the actual numbers of events in the various bins, but rather the corresponding expectation values, i.e. the μ_i are not in general integers. One could, for example, regard the true number of events in bin i as a random variable m_i with mean μ_i. Because of the limited resolution and efficiency, however, m_i is not directly observable, and it does not even enter the present formulation of the problem. Instead, we will construct estimators directly for the parameters μ_i.

For reasons of convenience one usually constructs a histogram of the observed values as well. Suppose that we begin with a sample of measured values of x, and that these are entered into a histogram with N bins, yielding $\mathbf{n} = (n_1, \ldots, n_N)$. These values could also be sample moments, Fourier coefficients, etc. In fact, the variable x could be multidimensional, containing not only a direct measurement of the true quantity of interest y, but also correlated quantities which provide additional information on y.

The number of bins N may in general be greater, less than, or equal to the number of bins M in the true histogram. Suppose the ith bin contains n_i entries, and that the total number of entries is $\sum_i n_i = n_{\text{tot}}$. It is often possible to regard the variables n_i as independent Poisson variables with expectation values ν_i. That is, for this model the probability to observe n_i entries in bin i is given by

$$P(n_i; \nu_i) = \frac{\nu_i^{n_i} e^{-\nu_i}}{n_i!}. \tag{11.3}$$

Since a sum of Poisson variables is itself a Poisson variable (cf. Section 10.4), n_{tot} will then follow a Poisson distribution with expectation value $\nu_{\text{tot}} = \sum_i \nu_i$. We may also consider the case where n_{tot} is regarded as a fixed parameter, and where the n_i follow a multinomial distribution. Whatever the distribution, we will call the expectation values

$$\nu_i = E[n_i]. \tag{11.4}$$

The form of the probability distribution for the data $\mathbf{n} = (n_1, \ldots, n_N)$ (Poisson, multinomial, etc.) will be needed in order to construct the likelihood function, used in unfolding methods based on maximum likelihood. Alternatively, we may be given the covariance matrix,

$$V_{ij} = \mathrm{cov}[n_i, n_j], \tag{11.5}$$

which is needed in methods based on least squares. We will assume that either the form of the probability law or the covariance matrix is known.

By using the law of total probability, (1.27), the expectation values $\nu_i = E[n_i]$ can be expressed as

$$
\begin{aligned}
\nu_i &= \mu_{\mathrm{tot}} \, P(\text{event observed in bin } i) \\[2mm]
&= \mu_{\mathrm{tot}} \int dy \, P \left(\begin{array}{c} \text{observed} \\ \text{in bin } i \end{array} \middle| \begin{array}{c} \text{true value } y \text{ and} \\ \text{event detected} \end{array} \right) \varepsilon(y) \, f_{\mathrm{true}}(y) \\[2mm]
&= \mu_{\mathrm{tot}} \int_{\mathrm{bin}\, i} dx \int dy \, s(x|y) \, \varepsilon(y) \, f_{\mathrm{true}}(y).
\end{aligned} \tag{11.6}
$$

Here $s(x|y)$ is the conditional p.d.f. for the measured value x given that the true value was y, and given that the event was observed somewhere, i.e. it is normalized such that $\int s(x|y)dx = 1$. We will call s the **resolution function** or in imaging applications the **point spread function**. One can also define a **response function**,

$$r(x|y) = s(x|y) \, \varepsilon(y), \tag{11.7}$$

which gives the probability to observe x, including the effect of limited efficiency, given that the true value was y. Note that this is not normalized as a conditional p.d.f. for x. One says that the true distribution is **folded** with the response function, and thus the task of estimating f_{true} is called **unfolding**.

Breaking the integral over y in equation (11.6) into a sum over bins and multiplying both numerator and denominator by μ_j, the expected number of entries to be observed in bin i becomes

$$
\begin{aligned}
\nu_i &= \sum_{j=1}^{M} \frac{\int_{\mathrm{bin}\, i} dx \int_{\mathrm{bin}\, j} dy \, s(x|y) \, \varepsilon(y) \, f_{\mathrm{true}}(y)}{(\mu_j / \mu_{\mathrm{tot}})} \mu_j \\[2mm]
&= \sum_{j=1}^{M} R_{ij} \, \mu_j,
\end{aligned} \tag{11.8}
$$

where the **response matrix** R is given by

$$R_{ij} = \frac{\int_{\text{bin } i} dx \int_{\text{bin } j} dy \, s(x|y) \, \varepsilon(y) \, f_{\text{true}}(y)}{\int_{\text{bin } j} dy \, f_{\text{true}}(y)}$$

$$= \frac{P(\text{observed in bin } i \text{ and true value in bin } j)}{P(\text{true value in bin } j)}$$

$$= P(\text{observed in bin } i \,|\, \text{true value in bin } j). \tag{11.9}$$

The response matrix element R_{ij} is thus the conditional probability that an event will be found with measured value x in bin i given that the true value y was in bin j. The effect of off-diagonal elements in R is to smear out any fine structure. A peak in the true histogram concentrated mainly in one bin will be observed over several bins. Two peaks separated by less than several bins will be merged into a single broad peak.

As can be seen from the first line of equation (11.9), the response matrix depends on the p.d.f. $f_{\text{true}}(y)$. This is a priori unknown, however, since the goal of the problem is to determine $f_{\text{true}}(y)$. If the bins of the unfolded histogram are small enough that $s(x|y)$ and $\varepsilon(y)$ are approximately constant over the bin of y, then the direct dependence on $f_{\text{true}}(y)$ cancels out. In the following we will assume that this approximation holds, and that the error in the response matrix due to any uncertainty in $f_{\text{true}}(y)$ can be neglected. In practice, the response matrix will be determined using whatever best approximation of $f_{\text{true}}(y)$ is available prior to carrying out the experiment.

Although $s(x|y)$ and $\varepsilon(y)$ are by construction independent of the probability that a given value y occurs (i.e. independent of $f_{\text{true}}(y)$), they are not in general completely model independent. The variable y may not be the only quantity that influences the probability to obtain a measured value x. For the example of beta decay where y represents the true and x the measured energy of the emitted electron, $s(x|y)$ and $\varepsilon(y)$ will depend in general on the angular distribution of the electrons (some parts of the detector may have better resolution than others), and different models of beta decay might predict different angular distributions.

In the following we will neglect this model dependence and simply assume that the resolution function $s(x|y)$ and efficiency $\varepsilon(y)$, and hence the response matrix R_{ij}, depend only on the measurement apparatus. We will assume in fact that R can be determined with negligible uncertainty both from the standpoint of model dependence as well as from that of detector response. In practice, R is determined either by means of calibration experiments where the true value y is known a priori, or by using a Monte Carlo simulation based on an understanding of the physical processes that take place in the detector. In real problems the model dependence may not be negligible, and the understanding of the detector itself is never perfect. Both must be treated as a possible sources of systematic error.

Note that the response matrix R_{ij} is not in general symmetric (nor even square), with the first index $i = 1, \ldots, N$ denoting the bin of the observed histogram and the second index $j = 1, \ldots, M$ referring to a bin of the true histogram. Summing over the first index and using $\int s(x|y)dx = 1$ gives

$$\sum_{i=1}^{N} R_{ij} = \sum_{i=1}^{N} \frac{\int_{\text{bin } i} dx \int_{\text{bin } j} dy\, s(x|y)\, \varepsilon(y)\, f_{\text{true}}(y)}{(\mu_j/\mu_{\text{tot}})}$$

$$= \frac{\int_{\text{bin } j} dy\, \varepsilon(y)\, f_{\text{true}}(y)}{\int_{\text{bin } j} f_{\text{true}}(y)\, dy}$$

$$\equiv \varepsilon_j, \tag{11.10}$$

i.e. one obtains the average value of the efficiency over bin j.

In addition to limited resolution and efficiency, one must also allow for the possibility that the measuring device produces a value when no true event of the type under study occurred, i.e. the measured 'value was caused by some **background** process. In the case of beta decay, this could be the result of spurious signals in the detector, the presence of radioactive nuclei in the sample other than the type under study, interactions due to particles coming from outside the apparatus such as cosmic rays, etc. Suppose that we have an expectation value β_i for the number of entries observed in bin i which originate from background processes. The relation (11.8) is then modified to be

$$\nu_i = \sum_{j=1}^{M} R_{ij}\, \mu_j + \beta_i. \tag{11.11}$$

Note that the β_i include the effects of limited resolution and efficiency of the detector. They will usually be determined either from calibration experiments or from a Monte Carlo simulation of both the background processes and the detector response. In the following we will assume that the values β_i are known, although in practice this will only be true to a given accuracy. The uncertainty in the background is thus a source of systematic error in the unfolded result.

To summarize, we have the following vector quantities (referred to also in a general sense as histograms or distributions):

(1) the true histogram (expectation values of true numbers of entries in each bin), $\boldsymbol{\mu} = (\mu_1, \ldots, \mu_M)$,
(2) the normalized true histogram (probabilities), $\mathbf{p} = (p_1, \ldots, p_M) = \boldsymbol{\mu}/\mu_{\text{tot}}$,
(3) the expectation values of the observed numbers of entries, $\boldsymbol{\nu} = (\nu_1, \ldots, \nu_N)$,
(4) the actual number of entries observed (the data), $\mathbf{n} = (n_1, \ldots, n_N)$,
(5) efficiencies $\boldsymbol{\varepsilon} = (\varepsilon_1, \ldots, \varepsilon_M)$, and
(6) expected background values $\boldsymbol{\beta} = (\beta_1, \ldots, \beta_N)$.

It is assumed either that we know the form of the probability distribution for the data $\mathbf{n}$, which will allow us to construct the likelihood function, or that we have the covariance matrix $V_{ij} = \text{cov}[n_i, n_j]$, which can be used to construct a χ^2 function. In addition we have the response matrix R_{ij}, where $i = 1, \ldots, N$ represents the bin of the observed histogram, and $j = 1, \ldots, M$ gives the bin of the true histogram. We will assume that R and β are known. The vectors μ, ν, β and the matrix R are related by

$$\nu = R\mu + \beta, \tag{11.12}$$

where μ, ν and β should be understood as column vectors in matrix equations. The goal is to construct estimators $\hat{\mu}$ for the true histogram, or estimators $\hat{\mathbf{p}}$ for the probabilities.

11.2 Inverting the response matrix

In this section we will examine a seemingly obvious method for constructing estimators for the true histogram μ, which, however, often leads to an unacceptable solution. Consider the case where the number of bins in the true and observed histograms are equal, $M = N$. For now we will assume that the matrix relation $\nu = R\mu + \beta$ can be inverted to give

$$\mu = R^{-1}(\nu - \beta). \tag{11.13}$$

An obvious choice for the estimators of ν is given by the corresponding data values,

$$\hat{\nu} = \mathbf{n}. \tag{11.14}$$

The estimators for the μ are then simply

$$\hat{\mu} = R^{-1}(\mathbf{n} - \beta). \tag{11.15}$$

One can easily show that this is, in fact, the solution obtained from maximizing the log-likelihood function,

$$\log L(\mu) = \sum_{i=1}^{N} \log P(n_i; \nu_i), \tag{11.16}$$

where $P(n_i; \nu_i)$ is a Poisson or binomial distribution. It is also the least squares solution, where one minimizes

$$\chi^2(\mu) = \sum_{i,j=1}^{N} (\nu_i - n_i)(V^{-1})_{ij}(\nu_j - n_j). \tag{11.17}$$

Note that $\log L(\mu)$ and $\chi^2(\mu)$ can be written as functions of μ or ν, since the relation $\nu = R\mu + \beta$ always holds. That is, when differentiating (11.16) or (11.17) with respect to μ_i one uses $\partial \nu_i / \partial \mu_j = R_{ij}$.

Before showing how the estimators constructed in this way can fail, it is interesting to investigate their bias and variance. The expectation value of $\hat{\mu}_j$ is given by

$$E[\hat{\mu}_j] = \sum_{i=1}^{N}(R^{-1})_{ji} \, E[n_i - \beta_i] = \sum_{i=1}^{N}(R^{-1})_{ji} \, (\nu_i - \beta_i)$$

$$= \mu_j, \tag{11.18}$$

so the estimators $\hat{\mu}_j$ are unbiased, since by assumption, $\hat{\nu}_i = n_i$ is unbiased. For the covariance matrix we find

$$\text{cov}[\hat{\mu}_i, \hat{\mu}_j] = \sum_{k,l=1}^{N}(R^{-1})_{ik} \, (R^{-1})_{jl} \, \text{cov}[n_k, n_l]$$

$$= \sum_{k=1}^{N}(R^{-1})_{ik} \, (R^{-1})_{jk} \, \nu_k, \tag{11.19}$$

where to obtain the last line we have used the covariance matrix for independent Poisson variables, $\text{cov}[n_k, n_l] = \delta_{kl}\nu_k$.

In the following we will use the notation $V_{ij} = \text{cov}[n_i, n_j]$ for the covariance matrix of the data, and $U_{ij} = \text{cov}[\hat{\mu}_i, \hat{\mu}_j]$ for that of the estimators of the true distribution. Equation (11.19) can then be written in matrix notation,

$$U = R^{-1} V \, (R^{-1})^{T}. \tag{11.20}$$

Consider now the example shown in Fig. 11.1. The original true distribution $\boldsymbol{\mu}$ is shown in Fig. 11.1(a), and the expectation values for the observed distribution $\boldsymbol{\nu}$ are shown in the histogram of Fig. 11.1(b).

The histogram $\boldsymbol{\nu}$ has been computed according to $\boldsymbol{\nu} = R\boldsymbol{\mu}$, i.e. the background $\boldsymbol{\beta}$ is taken to be zero. The response matrix R is based on a Gaussian resolution function with a standard deviation equal to 1.5 times the bin width, and the efficiencies ε_i are all taken to be unity. This results in a probability of approximately 26% for an event to remain in the bin where it was created, 21% for the event to migrate one bin, and 16% to migrate two or more bins.

Figure 11.1(c) shows the data $\mathbf{n} = (n_1, \ldots, n_N)$. These have been generated by the Monte Carlo method using Poisson distributions with the mean values ν_i from Fig. 11.1(b). Since the number of entries in each bin ranges from around 10^2 to 10^3, the relative statistical errors (ratio of standard deviation to mean value) for the n_i are in the range from 3 to 10%.

Figure 11.1(d) shows the estimates $\hat{\boldsymbol{\mu}}$ obtained from matrix inversion, equation (11.15). The error bars indicate the standard deviations for each bin. Far from achieving the 3–10% precision that we had for the n_i, the $\hat{\mu}_j$ oscillate

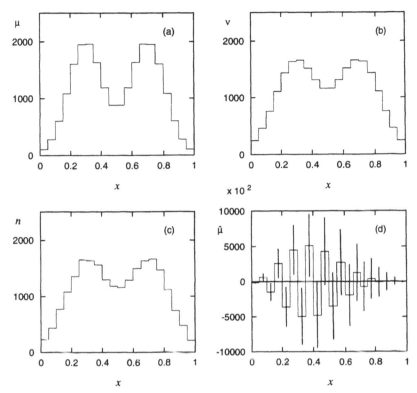

Fig. 11.1 (a) A hypothetical true histogram μ, (b) the histogram of expectation values $\nu = R\mu$, (c) the histogram of observed data **n**, and (d) the estimators $\hat{\mu}$ obtained from inversion of the response matrix.

wildly from bin to bin, and the error bars are as large as the estimated values themselves. (Notice the increased vertical scale on this plot.) The correlation coefficients for neighboring bins are close to -1.

The reason for the catastrophic failure stems from the fact that we do not have the expectation values ν; if we did, we could simply compute $\mu = R^{-1}\nu$. Rather, we only have the data **n**, which are random variables and hence subject to statistical fluctuations. Recall that the effect of the response matrix is to smear out any fine structure. If there had been peaks close together in μ, then although these would be merged together in ν, there would still remain a certain residual fine structure. Upon applying R^{-1} to ν, this remnant of the original structure would be restored. The data **n** have indeed statistical fluctuations from bin to bin, and this leads to the same qualitative result as would a residual fine structure in ν. Namely, the unfolded result is given a large amount of fine structure, as is evident in Fig. 11.1(d).

It is interesting to compare the covariance matrix U (11.19) with that given by the RCF inequality (cf. Section 6.6); this gives the smallest possible variance for any choice of estimator. For this we will regard the n_i as independent Poisson variables with mean values ν_i. The log-likelihood function is thus

$$\log L(\boldsymbol{\mu}) = \sum_{i=1}^{N} \log P(n_i; \nu_i) = \sum_{i=1}^{N} \log \left(\frac{\nu_i^{n_i} e^{-\nu_i}}{n_i!} \right). \tag{11.21}$$

Dropping additive terms that do not depend on $\boldsymbol{\mu}$ gives

$$\log L(\boldsymbol{\mu}) = \sum_{i=1}^{N} (n_i \log \nu_i - \nu_i). \tag{11.22}$$

One can check that by setting the derivatives of $\log L$ with respect to the components of $\boldsymbol{\mu}$ equal to zero,

$$\frac{\partial \log L}{\partial \mu_k} = \sum_{i=1}^{N} \frac{\partial \log L}{\partial \nu_i} \frac{\partial \nu_i}{\partial \mu_k} = \sum_{i=1}^{N} \left(\frac{n_i}{\nu_i} - 1 \right) R_{ik} = 0, \tag{11.23}$$

one obtains in fact the same estimators, $\hat{\boldsymbol{\nu}} = \mathbf{n}$, as we have seen previously. Differentiating one more time gives

$$\frac{\partial^2 \log L}{\partial \mu_k \, \partial \mu_l} = -\sum_{i=1}^{N} \frac{n_i \, R_{ik} \, R_{il}}{\nu_i^2}. \tag{11.24}$$

The RCF bound for the inverse covariance matrix for the case of zero bias (equation (6.19)) is therefore

$$
\begin{aligned}
(U^{-1})_{kl} &= -E \left[\frac{\partial^2 \log L}{\partial \mu_k \, \partial \mu_l} \right] \\[2mm]
&= \sum_{i=1}^{N} \frac{E[n_i] \, R_{ik} \, R_{il}}{\nu_i^2} \\[2mm]
&= \sum_{i=1}^{N} \frac{R_{ik} \, R_{il}}{\nu_i}.
\end{aligned} \tag{11.25}
$$

By multiplying both sides of the equation once by U, twice by R^{-1}, and summing over the appropriate indices, one can solve (11.25) for the RCF bound for the covariance matrix,

$$U_{ij} = \sum_{k=1}^{N} (R^{-1})_{ik} \, (R^{-1})_{jk} \, \nu_k. \tag{11.26}$$

This is the same as the result of the exact calculation (11.19), so we see that the maximum likelihood solution is both unbiased and efficient, i.e. it has the smallest possible variance for an estimator with zero bias. We would obtain the same result using the method of least squares; in that case, unbiased and efficient estimators are guaranteed by the Gauss–Markov theorem.

Although the solution in Fig. 11.1(d) bears little resemblance to the true distribution, it has certain desirable properties. It is simple to construct, has zero bias, and the variance is equal to the RCF bound. In order to be of use, however, the correlations must be taken into account. For example, one can test the compatibility of the estimators $\hat{\boldsymbol{\mu}}$ with a hypothesis $\boldsymbol{\mu}_0$ by constructing a χ^2 statistic,

$$\chi^2 = (\hat{\boldsymbol{\mu}} - \boldsymbol{\mu}_0)^T \, U^{-1} \, (\hat{\boldsymbol{\mu}} - \boldsymbol{\mu}_0), \tag{11.27}$$

which uses the full covariance matrix U of the estimators. This test would be meaningless if the χ^2 were to be computed with only the diagonal elements of U. We should also note that although the variances are extremely large in the example shown here, they would be significantly smaller if the bins are made large compared to the width of the resolution function.

Regardless of its drawbacks, response-matrix inversion indicates some important lessons and provides a starting point for other methods. Since the inverse-matrix solution has zero bias and minimum variance as given by the RCF inequality, any reduction in variance can only be achieved by introducing a bias. The art of unfolding consists of constructing biased estimators $\hat{\boldsymbol{\mu}}$ such that the bias will be small if our prior beliefs, usually some assumptions concerning smoothness, are in fact correct. Roughly speaking, the goal is to find an optimal trade-off between bias and variance, although we will see in Section 11.7 that there is a certain arbitrariness in determining how this optimum is achieved.

The need to incorporate prior knowledge suggests using the Bayesian approach, where the a priori probabilities are combined with the data to yield a posteriori probabilities for the true distribution (cf. Sections 1.2, 6.13). This is a common starting point in the literature on unfolding. It suffers from the difficulty, however, that prior knowledge is often of a complicated or qualitative nature and is thus difficult to express in terms of prior probabilities. The fact that prior beliefs are inherently subjective is not a real disadvantage here; in the classical approach as well there is a certain subjectivity as to how one chooses a biased estimator. In the following we will mainly follow classical statistics, using bias and variance as the criteria by which to judge the quality of a solution, while pointing out the connections with the Bayesian techniques wherever possible.

As a final remark on matrix inversion, we can consider the case where the number of bins M in the unfolded histogram is not equal to the number of measured bins N. For $M > N$, the system of equations (11.12), $\boldsymbol{\nu} = R\boldsymbol{\mu} + \boldsymbol{\beta}$, is underdetermined, and the solution is not unique. The methods presented in Section 11.4 can be used to select a solution as the estimator $\hat{\boldsymbol{\mu}}$. For $M < N$, (11.12) is overdetermined, and an exact solution does not exist in general. An approximate solution can be constructed using, for example, the methods of maximum likelihood or of least squares, i.e. the problem is equivalent to parameter estimation as discussed in Chapters 5–8. If M is large, then correlations between the estimators can lead to large variances. In such a case it may be desirable to reduce the variances, at the cost of introducing bias, by using one of the regularization methods of Section 11.4.

11.3 The method of correction factors

Consider the case where the bins of the true distribution μ are taken to be the same as those of the data $\mathbf{n}$. One of the simplest and perhaps most commonly used techniques is to take as the estimator for μ_i

$$\hat{\mu}_i = C_i(n_i - \beta_i), \tag{11.28}$$

where β_i is the expected background and C_i is a multiplicative **correction factor**. The correction factors can be determined using a Monte Carlo program which includes both a model of the process under study as well as a simulation of the measuring apparatus. The factors C_i are determined by running the Monte Carlo program once with and once without the detector simulation, yielding model predictions for the observed and true values of each bin, ν_i^{MC} and μ_i^{MC}. Here ν^{MC} refers to the signal process only, i.e. background is not included. The correction factor is then simply the ratio,

$$C_i = \frac{\mu_i^{\mathrm{MC}}}{\nu_i^{\mathrm{MC}}}. \tag{11.29}$$

For now we will assume that it is possible to generate enough Monte Carlo data so that the statistical errors in the correction factors are negligible. If this is not the case, the uncertainties in the C_i can be incorporated into those of the estimates $\hat{\mu}_i$ by the usual procedure of error propagation.

If the effects of resolution are negligible, then the response matrix is diagonal, i.e. $R_{ij} = \delta_{ij}\varepsilon_j$, and therefore one has

$$\nu_i^{\mathrm{sig}} = \nu_i - \beta_i = \varepsilon_i\mu_i, \tag{11.30}$$

where ν_i^{sig} is the expected number of entries in bin i without background. Thus the correction factors become simply $C_i = 1/\varepsilon_i$, so that $1/C_i$ plays the role of a generalized efficiency. When one has off-diagonal terms in the response matrix, however, the values of $1/C_i$ can be greater than unity. That is, because of migrations between bins, it is possible to find more entries in a given bin than the number of true entries actually created there.

The expectation value of the estimator $\hat{\mu}_i$ is

$$E[\hat{\mu}_i] \quad = \quad C_i\, E[n_i - \beta_i] = C_i(\nu_i - \beta_i) = \frac{\mu_i^{\text{MC}}}{\nu_i^{\text{MC}}}\, \nu_i^{\text{sig}}$$

$$= \quad \left(\frac{\mu_i^{\text{MC}}}{\nu_i^{\text{MC}}} - \frac{\mu_i}{\nu_i^{\text{sig}}} \right) \nu_i^{\text{sig}} + \mu_i. \qquad (11.31)$$

The estimator $\hat{\mu}_i$ thus has a bias which is only zero if the ratios μ_i/ν_i^{sig} are the same for the Monte Carlo model and for the real experiment.

The covariance matrix for the estimators is given by

$$\text{cov}[\hat{\mu}_i, \hat{\mu}_j] \quad = \quad C_i^2\, \text{cov}[n_i, n_j]$$

$$= \quad C_i^2\, \delta_{ij}\, \nu_i. \qquad (11.32)$$

The last line uses the covariance matrix for the case where the n_i are independent Poisson variables with expectation values ν_i. For many practical problems, the C_i are of order unity, and thus the variances of the estimates $\hat{\mu}_i$ are approximately the same as what one would achieve with perfect resolution. In addition, the technique is simple to implement, not even requiring a matrix inversion. The price that one pays is the bias,

$$b_i = \left(\frac{\mu_i^{\text{MC}}}{\nu_i^{\text{MC}}} - \frac{\mu_i}{\nu_i^{\text{sig}}} \right) \nu_i^{\text{sig}}. \qquad (11.33)$$

A rough estimate of the systematic uncertainty due to this bias can be obtained by computing the correction factors with different Monte Carlo models. Clearly a better model leads to a smaller bias, and therefore it is often recommended that the estimated distribution $\hat{\mu}$ be used to tune the Monte Carlo, i.e. by adjusting its parameters to improve the agreement between ν^{MC} and the background subtracted data $\mathbf{n} - \beta$. One can then iterate the procedure and obtain improved correction factors from the tuned model.

A danger in the method of correction factors is that the bias often pulls the estimates $\hat{\mu}$ towards the model prediction μ^{MC}. This complicates the task of testing the model, which may have been the purpose of carrying out the measurement in the first place. In such cases one must ensure that the uncertainty in the unfolded result due to the model dependence of the correction factors is taken into account in the estimated systematic errors, and that these are incorporated into any model tests.

11.4 General strategy of regularized unfolding

Although the method of correction factors is simple and widely practiced, it has a number of disadvantages, primarily related to the model dependence of the result. An alternative approach is to impose in some way a measure of smoothness on the estimators for the true histogram $\boldsymbol{\mu}$. This is known as **regularization** of the unfolded distribution.

As a starting point, let us return to the oscillating solution of Section 11.2 obtained from inversion of the response matrix. This estimate for $\boldsymbol{\mu}$ is characterized by a certain maximum value of the log-likelihood function $\log L_{\max}$, or a minimum value of the χ^2. In the following we will usually refer only to the log-likelihood function; the corresponding relations using χ^2 can be obtained by the replacement $\log L = -\chi^2/2$.

One can consider a certain region of $\boldsymbol{\mu}$-space around the maximum likelihood (or least squares) solution as representing acceptable solutions, in the sense that they have an acceptable level of agreement between the predicted expectation values $\boldsymbol{\nu}$ and the data $\mathbf{n}$. The extent of this region can be defined by requiring that $\log L$ stay within some limit of its maximum value. That is, one determines the acceptable region of $\boldsymbol{\mu}$-space by

$$\log L(\boldsymbol{\mu}) \geq \log L_{\max} - \Delta \log L \tag{11.34}$$

or for the case of least squares,

$$\chi^2(\boldsymbol{\mu}) \leq \chi^2_{\min} + \Delta\chi^2 \tag{11.35}$$

for appropriately chosen $\Delta \log L$ or $\Delta\chi^2$. The values of $\Delta \log L$ or $\Delta\chi^2$ will determine the trade-off between bias and variance achieved in the unfolded histogram; we will return to this point in detail in Section 11.7.

In addition to the acceptability of the solution, we need to define a measure of its smoothness by introducing a function $S(\boldsymbol{\mu})$, called the **regularization function**. Several possible forms for $S(\boldsymbol{\mu})$ will be discussed in the next section. The general strategy is to choose the solution with the highest degree of smoothness out of the acceptable solutions determined by the inequalities (11.34) or (11.35).

Maximizing the regularization function $S(\boldsymbol{\mu})$ with the constraint that $\log L(\boldsymbol{\mu})$ remain equal to $\log L_{\max} - \Delta \log L$ is equivalent to maximizing the quantity

$$\alpha \left[\log L(\boldsymbol{\mu}) - (\log L_{\max} - \Delta \log L)\right] + S(\boldsymbol{\mu}) \tag{11.36}$$

with respect to both $\boldsymbol{\mu}$ and α. Here α is a Lagrange multiplier called the **regularization parameter**, which can be chosen to correspond to a specific value of $\Delta \log L$. For a given α, the solution is thus determined by finding the maximum of a weighted combination of $\log L$ and the $S(\boldsymbol{\mu})$,

$$\Phi(\boldsymbol{\mu}) = \alpha \, \log L(\boldsymbol{\mu}) + S(\boldsymbol{\mu}). \tag{11.37}$$

Setting $\alpha = 0$ leads to the smoothest distribution possible; this ignores completely the data $\mathbf{n}$. A very large α leads to the oscillating solution from inversion of the response matrix, corresponding to having the likelihood function equal to its maximum value.

In order for the prescription of maximizing $\Phi(\boldsymbol{\mu})$ to be in fact equivalent to the general strategy stated above, the surfaces of constant $\log L(\boldsymbol{\mu})$ and $S(\boldsymbol{\mu})$

must be sufficiently well behaved; in the following we will assume this to be the case. In particular, they should not change from convex to concave or have a complicated topology such that multiple local maxima exist.

Recall that we can write $\log L$ and S as functions of $\boldsymbol{\mu}$ or $\boldsymbol{\nu}$, since the relation $\boldsymbol{\nu} = R\boldsymbol{\mu} + \boldsymbol{\beta}$ always holds. In a similar way, we will always take the relation

$$\hat{\boldsymbol{\nu}} = R\hat{\boldsymbol{\mu}} + \boldsymbol{\beta} \tag{11.38}$$

to define the estimators for $\boldsymbol{\nu}$; knowing these is equivalent to knowing the estimators $\hat{\boldsymbol{\mu}}$. Note, however, that in contrast to the method of Section 11.2, we will no longer have $\hat{\boldsymbol{\nu}} = \mathbf{n}$. It should also be kept in mind that $\mu_{\text{tot}} = \sum_j \mu_j$ and $\nu_{\text{tot}} = \sum_i \nu_i = \sum_{i,j} R_{ij}\mu_j$ are also functions of $\boldsymbol{\mu}$.

Here we will only consider estimators $\hat{\boldsymbol{\mu}}$ for which the estimated total number of events $\hat{\nu}_{\text{tot}}$ is equal to the number actually observed,

$$\hat{\nu}_{\text{tot}} = \sum_{i=1}^{N} \hat{\nu}_i = \sum_{i=1}^{N}\sum_{j=1}^{M} R_{ij}\,\hat{\mu}_j + \beta_i = n_{\text{tot}}. \tag{11.39}$$

This condition is not in general fulfilled automatically. It can be imposed by modifying equation (11.37) to read

$$\varphi(\boldsymbol{\mu}, \lambda) = \alpha \log L(\boldsymbol{\mu}) + S(\boldsymbol{\mu}) + \lambda \left[n_{\text{tot}} - \sum_{i=1}^{N} \nu_i \right], \tag{11.40}$$

where λ is a Lagrange multiplier. Setting $\partial\varphi/\partial\lambda = 0$ then leads to $\sum_i \nu_i = n_{\text{tot}}$.

As a technical aside, note that it does not matter whether the regularization parameter α is attached to the regularization function $S(\boldsymbol{\mu})$ (as it is in many references) or with the likelihood function. In the particular numerical implementation given in Section 11.9, it is more convenient to associate α with the likelihood.

11.5 Regularization functions

11.5.1 Tikhonov regularization

A commonly used measure of smoothness is the mean value of the square of some derivative of the true distribution. This technique was suggested independently by Phillips [Phi62] and Tikhonov [Tik63, Tik77], and is usually called **Tikhonov regularization**. If we consider the p.d.f. $f_{\text{true}}(y)$ before being discretized as a histogram, then the regularization function is

$$S[f_{\text{true}}(y)] = - \int \left(\frac{d^k f_{\text{true}}(y)}{dy^k} \right)^2 dy, \tag{11.41}$$

where the integration is over all allowed values of y. The minus sign comes from the convention taken here that we maximize φ as defined by (11.40). That is, greater S corresponds to more smoothness. (Equivalently one can of course minimize a combination of regularization and log-likelihood functions with the opposite sign; this convention as well is often encountered in the literature.)

In principle, a linear combination of terms with different derivatives could be used; in practice, one value of k is usually chosen. When $f_{\text{true}}(y)$ is represented as a histogram, the derivatives are replaced by finite differences. For equal bin widths, one can use for $k = 1$ (cf. [Pre92])

$$S(\boldsymbol{\mu}) = -\sum_{i=1}^{M-1} (\mu_i - \mu_{i+1})^2, \tag{11.42}$$

for $k = 2$

$$S(\boldsymbol{\mu}) = -\sum_{i=1}^{M-2} (-\mu_i + 2\mu_{i+1} - \mu_{i+2})^2, \tag{11.43}$$

or for $k = 3$

$$S(\boldsymbol{\mu}) = -\sum_{i=1}^{M-3} (-\mu_i + 3\mu_{i+1} - 3\mu_{i+2} + \mu_{i+3})^2. \tag{11.44}$$

A common choice for the derivative is $k = 2$, so that $S(\boldsymbol{\mu})$ is related to the average curvature.

If the bin widths Δy_i are all equal, then they can be ignored in (11.42)–(11.44). This would only give a constant of proportionality, and can be effectively absorbed into the regularization parameter α. If the Δy_i are not all equal, then this can be included in the finite differences in a straightforward manner. For $k = 2$, for example, one can assume a parabolic form for $f_{\text{true}}(y)$ within each group of three adjacent bins,

$$f_i(y) = a_{0i} + a_{1i} y + a_{2i} y^2. \tag{11.45}$$

There are $M - 2$ such groups, centered around bins $i = 2, \ldots, M - 1$. The coefficients can be determined in each group by setting the integrals of $f_i(y)$ over bins $i - 1$, i and $i + 1$ equal to the corresponding values of μ_{i-1}, μ_i and μ_{i+1}. The second derivative for the group centered around bin i is then $f_i'' = 2a_{2i}$, and the regularization function can thus be taken to be

$$S(\boldsymbol{\mu}) = -\sum_{i=2}^{M-1} f_i''^2 \, \Delta y_i. \tag{11.46}$$

Note that the second derivative cannot be determined in the first and last bins. Here they are not included in the sum (11.46), i.e. they are taken to be zero; alternatively one could set them equal to the values obtained in bins 2 and $M - 1$.

For any value of the derivative k and regardless of the bin widths, the functions $S(\boldsymbol{\mu})$ given above can be expressed as

$$S(\boldsymbol{\mu}) = - \sum_{i,j=1}^{M} G_{ij}\, \mu_i\, \mu_j = -\boldsymbol{\mu}^T G\, \boldsymbol{\mu}, \qquad (11.47)$$

where G is a symmetric matrix of constants. For $k = 2$ with equal bin widths (11.43), for example, G is given by

$$\left.\begin{array}{l} G_{ii} = 6 \\ G_{i,i\pm1} = G_{i\pm1,i} = -4 \\ G_{i,i\pm2} = G_{i\pm2,i} = 1 \end{array}\right\} \quad 3 \le i \le M - 2, \\ G_{11} = G_{MM} = 1, \\ G_{22} = G_{M-1,M-1} = 5, \\ G_{12} = G_{21} = G_{M,M-1} = G_{M-1,M} = -2, \qquad (11.48)$$

with all other G_{ij} equal to zero.

In order to obtain the estimators and their covariance matrix (Section 11.6), we will need the first and second derivatives of S. These are

$$\frac{\partial S}{\partial \mu_i} = -2 \sum_{j=1}^{M} G_{ij}\, \mu_j \qquad (11.49)$$

and

$$\frac{\partial^2 S}{\partial \mu_i \partial \mu_j} = -2\, G_{ij}. \qquad (11.50)$$

Tikhonov regularization using $k = 2$ has been widely applied in particle physics for the unfolding of structure functions (distributions of kinematic variables in lepton–nucleon scattering). Further descriptions can be found in [Blo85, Höc96, Roe92, Zec95].

11.5.2 Regularization functions based on entropy

Another commonly used regularization function is based on the **entropy** H of a probability distribution $\mathbf{p} = (p_1, \dots, p_M)$, defined as [Sha48]

$$H = - \sum_{i=1}^{M} p_i \log p_i. \qquad (11.51)$$

The idea here is to interpret the entropy as a measure of the smoothness of a histogram $\boldsymbol{\mu} = (\mu_1, \dots, \mu_M)$, and to use

$$S(\boldsymbol{\mu}) = H(\boldsymbol{\mu}) = - \sum_{i=1}^{M} \frac{\mu_i}{\mu_{\text{tot}}} \log \frac{\mu_i}{\mu_{\text{tot}}} \qquad (11.52)$$

as a regularization function. Estimators based on (11.52) are said to be constructed according to the **principle of maximum entropy** or **MaxEnt**. To see how entropy is related to smoothness, consider the number of ways in which a particular histogram $\boldsymbol{\mu} = (\mu_1, \ldots, \mu_M)$ can be constructed out of μ_{tot} entries (here the values μ_j are integers). This is given by

$$\Omega(\boldsymbol{\mu}) = \frac{\mu_{\text{tot}}!}{\mu_1! \, \mu_2! \ldots \mu_M!}. \tag{11.53}$$

(Recall that the same factor appears in the multinomial distribution (2.6).) By taking the logarithm of (11.53) and using Stirling's approximation, $\log n! \approx n(\log n - 1)$, valid for large n, one obtains

$$
\begin{aligned}
\log \Omega &\approx \mu_{\text{tot}} (\log \mu_{\text{tot}} - 1) - \sum_{i=1}^{M} \mu_i (\log \mu_i - 1) \\
&= -\sum_{i=1}^{M} \mu_i \log \frac{\mu_i}{\mu_{\text{tot}}} \\
&= \mu_{\text{tot}} \, S(\boldsymbol{\mu}).
\end{aligned}
\tag{11.54}
$$

We will use equation (11.54) to generalize $\log \Omega$ to the case where the μ_i are not integers.

If all of the events are concentrated in a single bin, i.e. the histogram has the minimum degree of smoothness, then there is only one way of arranging them, and hence the entropy is also a minimum. At the other extreme, one can show that the entropy is maximum for the case where all μ_i are equal, i.e. the histogram corresponds to a uniform distribution. To maximize H with the constraint $\sum_i p_i = 1$, a Lagrange multiplier can be used.

For later reference, we list here the first and second derivatives of the entropy-based $S(\boldsymbol{\mu})$:

$$\frac{\partial S}{\partial \mu_i} = -\frac{1}{\mu_{\text{tot}}} \log \frac{\mu_i}{\mu_{\text{tot}}} - \frac{S(\boldsymbol{\mu})}{\mu_{\text{tot}}} \tag{11.55}$$

and

$$\frac{\partial^2 S}{\partial \mu_i \partial \mu_j} = \frac{1}{\mu_{\text{tot}}^2} \left[1 - \frac{\delta_{ij} \, \mu_{\text{tot}}}{\mu_i} + \log \left(\frac{\mu_i \mu_j}{\mu_{\text{tot}}^2} \right) + 2S(\boldsymbol{\mu}) \right]. \tag{11.56}$$

11.5.3 Bayesian motivation for the use of entropy

In much of the literature on unfolding problems, the principle of maximum entropy is developed in the framework of Bayesian statistics. (See, for example, [Siv96, Jay86, Pre92].) This approach to unfolding runs into difficulties, however, as we will see below. It is nevertheless interesting to compare Bayesian MaxEnt with the classical methods of the previous section.

In the Bayesian approach, the values μ are treated as random variables in the sense of subjective probability (cf. Section 1.2), and the joint probability density $f(\mu|\mathbf{n})$ represents the degree of belief that the true histogram is given by μ. To update our knowledge about μ in light of the data $\mathbf{n}$, we use Bayes' theorem,

$$f(\mu|\mathbf{n}) \propto L(\mathbf{n}|\mu) \, \pi(\mu), \tag{11.57}$$

where $L(\mathbf{n}|\mu)$ is the likelihood function (the conditional probability for the data $\mathbf{n}$ given μ) multiplied by the prior density $\pi(\mu)$. The prior density represents our knowledge about μ before seeing the data $\mathbf{n}$.

Here we will regard the total number of events μ_{tot} as an integer. This is in contrast to the classical approach, where μ_{tot} represents an expectation value of an integer random variable, and thus is not necessarily an integer itself. Suppose we have no prior knowledge about how these μ_{tot} entries are distributed in the histogram. One can then argue that by symmetry, each of the possible ways of placing μ_{tot} entries into M bins is equally likely. The probability for a certain histogram $(\mu_1, \ldots, \mu_M)$ therefore should be, in the absence of any other prior information, proportional to the number of ways in which it can be made; this is just the number Ω given by equation (11.53). The total number of ways of distributing the entries $\Omega(\mu)$ is thus interpreted as the prior probability $\pi(\mu)$,

$$
\begin{aligned}
\pi(\mu) \;=\; \Omega(\mu) &= \frac{\mu_{\text{tot}}!}{\mu_1! \, \mu_2! \, \ldots \, \mu_M!} \\[2mm]
&= \exp(\mu_{\text{tot}} \, H),
\end{aligned}
\tag{11.58}
$$

where H is the entropy given by equation (11.51).

From the strict Bayesian standpoint, the job is finished when we have determined $f(\mu|\mathbf{n})$. It is not practical to report $f(\mu|\mathbf{n})$ completely, however, since this is a function of as many variables as there are bins M in the unfolded distribution. Therefore some way of summarizing it must be found; to do this one typically selects a single vector $\hat{\mu}$ as the Bayesian estimator. The usual choice is the μ for which the probability $f(\mu|\mathbf{n})$, or equivalently its logarithm, is a maximum. According to equation (11.57), this is determined by maximizing

$$
\begin{aligned}
\log f(\mu|\mathbf{n}) \;\propto\; &\log L(\mu|\mathbf{n}) + \log \pi(\mu) \\[1mm]
=\; &\log L(\mu|\mathbf{n}) + \mu_{\text{tot}} H(\mu) \\[1mm]
=\; &\log L(\mu|\mathbf{n}) + \mu_{\text{tot}} H(\mu).
\end{aligned}
\tag{11.59}
$$

The Bayesian prescription thus corresponds to using a regularization function

$$S(\mu) = \mu_{\text{tot}} H(\mu) = -\sum_{i=1}^{M} \mu_i \, \log \frac{\mu_i}{\mu_{\text{tot}}}. \tag{11.60}$$

Furthermore, the regularization parameter α is no longer an arbitrary factor but is set equal to 1. If all of the efficiencies ε_i are equal, then the requirement $\nu_{\text{tot}} = n_{\text{tot}}$ also implies that μ_{tot} is constant. This is then equivalent to using the previous regularization function $S(\boldsymbol{\mu}) = H$ with $\alpha = 1/\mu_{\text{tot}}$.

If the efficiencies are not all equal, however, then constant ν_{tot} does not imply constant μ_{tot}, and as a result, the distribution of maximum $S(\boldsymbol{\mu}) = \mu_{\text{tot}} H(\boldsymbol{\mu})$ is no longer uniform. This is because S can increase simply by increasing μ_{tot}, and thus in the distribution of maximum S, bins with low efficiency are enhanced. In this case, then, using H and $\mu_{\text{tot}} H$ as regularization functions will lead to somewhat different results, although the difference is in practice not great if the efficiencies are of the same order of magnitude. In any event, $S = H$ is easier to justify as a measure of smoothness, since the distribution of maximum H is always uniform.

We will see in Section 11.9 that the Bayesian estimator (11.59) gives too much weight to the entropy term (see Fig. 11.3(a) and [Ski86]). From the classical point of view one would say that it does not represent a good trade-off between bias and variance, having an unreasonably large bias. One can modify the Bayesian interpretation by replacing μ_{tot} in (11.59) by an effective number of events μ_{eff}, which can be adjusted to be smaller than μ_{tot}. The estimator is then given by the maximum of

$$\log L(\boldsymbol{\mu}|\mathbf{n}) + \mu_{\text{eff}} H(\boldsymbol{\mu}). \tag{11.61}$$

This is equivalent to using $S(\boldsymbol{\mu}) = H(\boldsymbol{\mu})$ as before, and the parameter μ_{eff} plays the role of the regularization parameter.

The problem with the original Bayesian solution stems from our use of $\Omega(\boldsymbol{\mu})$ as the prior density. From either the Bayesian or classical points of view, the quantities $\mathbf{p} = \boldsymbol{\mu}/\mu_{\text{tot}}$ are given by some set of unknown, constant numbers, e.g. the electron energy distribution of specific type of beta decay. In either case, our prior knowledge about the *form* of the distribution (i.e. about $\mathbf{p}$, not $\boldsymbol{\mu}$) should be independent of the number of observations in the data sample that we obtain. This points to a fundamental problem in using $\pi(\boldsymbol{\mu}) = \Omega(\boldsymbol{\mu})$, since this becomes increasingly concentrated about a uniform distribution (i.e. all p_i equal) as μ_{tot} increases.

It is often the case that we have indeed some prior beliefs about the form of the distribution $\mathbf{p}$, but that these are difficult to quantify. We could say, for example, that distributions with large amounts of structure are a priori unlikely, since it may be difficult to imagine a physical theory predicting something with lots of peaks. On the other hand, a completely flat distribution may not seem very physical either, so $\Omega(\boldsymbol{\mu})$ does not really reflect our prior beliefs. Because of these difficulties with the interpretation of $\Omega(\boldsymbol{\mu})$ as a prior p.d.f., we will stay with the classical approach here, and simply regard the entropy as one of the possible regularization functions.

11.5.4 Regularization function based on cross-entropy

Recall that the distribution of maximum entropy is flat, and thus the bias introduced into the estimators $\hat{\mu}$ will tend to pull the result towards a more uniform distribution. Suppose we know a distribution $\mathbf{q} = (q_1, \ldots, q_M)$ that we regard as the most likely a priori shape for the true distribution $\mathbf{p} = \mu/\mu_{tot}$. We will call $\mathbf{q}$ the **reference distribution**. Suppose that we do not know how to quantify our degree of belief in $\mathbf{q}$, however, and hence we do not have a prior density $\pi(\mu)$ for use with Bayes' theorem. That is, $\mathbf{q}$ represents the normalized histogram μ/μ_{tot} for which the prior density is a maximum, but it does not specify the entire prior density.

In this case, the regularization function can be taken as

$$S(\mu) = K(\mathbf{p}; \mathbf{q}), \tag{11.62}$$

where $K(\mathbf{p}; \mathbf{q})$ is called the **cross-entropy** [Kul64] or **Shannon–Jaynes entropy** [Jay68], defined as

$$K(\mathbf{p}; \mathbf{q}) = -\sum_{i=1}^{M} p_i \log \frac{p_i}{M q_i}. \tag{11.63}$$

The cross-entropy is often defined without the factor of M, and also without the minus sign, in which case the principle of maximum entropy becomes the principle of minimum cross-entropy. We will keep the minus sign so as to maintain the similarity between $K(\mathbf{p}; \mathbf{q})$ and the Shannon entropy $H(\mathbf{p})$ (11.51). Note that $K(\mathbf{p}; \mathbf{q}) = H(\mathbf{p})$ when the reference distribution is uniform, i.e. $q_i = 1/M$ for all i.

One can easily show that the cross-entropy $K(\mathbf{p}; \mathbf{q})$ is a maximum when the probabilities $\mathbf{p}$ are equal to those of the reference distribution $\mathbf{q}$. The effect of using the regularization function (11.62) is that the bias of the estimators $\hat{\mu}$ will be zero (or small) if the true distribution is equal (or close) to the reference distribution.

11.6 Variance and bias of the estimators

The estimators $\hat{\mu}$ are functions of the data $\mathbf{n}$, and are hence themselves random variables. In order to obtain the covariance matrix $U_{ij} = \text{cov}[\hat{\mu}_i, \hat{\mu}_j]$, we can calculate an approximate expression for $\hat{\mu}$ as a function of $\mathbf{n}$, and then use the error propagation formula (1.54) to relate U to the covariance matrix for the data, $V_{ij} = \text{cov}[n_i, n_j]$.

The estimators $\hat{\mu}$ are found by maximizing the function $\varphi(\mu, \lambda)$ given by (11.40), which uses a given log-likelihood or χ^2 function and some form of the regularization function $S(\mu)$ (Tikhonov, entropy, etc.). The estimators $\hat{\mu}$ and the Lagrange multiplier λ are thus solutions to the system of $M + 1$ equations

$$F_i(\mu, \lambda, \mathbf{n}) = 0, \quad i = 1, \ldots, M + 1, \tag{11.64}$$

where

$$F_i(\boldsymbol{\mu}, \lambda, \mathbf{n}) = \begin{cases} \frac{\partial \varphi}{\partial \mu_i} & i = 1, \dots, M, \\ \frac{\partial \varphi}{\partial \lambda} & i = M + 1. \end{cases} \tag{11.65}$$

Suppose the data actually obtained are given by the vector $\tilde{\mathbf{n}}$, the corresponding estimates are $\tilde{\boldsymbol{\mu}} = \hat{\boldsymbol{\mu}}(\tilde{\mathbf{n}})$, and the Lagrange multiplier λ has the value $\tilde{\lambda}$. We would like to know how $\hat{\boldsymbol{\mu}}$ and λ would change if the data were given by some different values $\mathbf{n}$. Expanding the functions $F_i(\boldsymbol{\mu}, \lambda, \mathbf{n})$ to first order in a Taylor series about the values $\tilde{\boldsymbol{\mu}}$, $\tilde{\lambda}$ and $\tilde{\mathbf{n}}$ gives

$$\begin{aligned} F_i(\boldsymbol{\mu}, \lambda, \mathbf{n}) &\approx F_i(\tilde{\boldsymbol{\mu}}, \tilde{\lambda}, \tilde{\mathbf{n}}) + \sum_{j=1}^{M} \left[\frac{\partial F_i}{\partial \mu_j} \right]_{\tilde{\boldsymbol{\mu}}, \tilde{\lambda}, \tilde{\mathbf{n}}} (\mu_j - \tilde{\mu}_j) \\ &+ \left[\frac{\partial F_i}{\partial \lambda} \right]_{\tilde{\boldsymbol{\mu}}, \tilde{\lambda}, \tilde{\mathbf{n}}} (\lambda - \tilde{\lambda}) + \sum_{j=1}^{N} \left[\frac{\partial F_i}{\partial n_j} \right]_{\tilde{\boldsymbol{\mu}}, \tilde{\lambda}, \tilde{\mathbf{n}}} (n_j - \tilde{n}_j). \end{aligned} \tag{11.66}$$

The first term $F_i(\tilde{\boldsymbol{\mu}}, \tilde{\lambda}, \tilde{\mathbf{n}})$ as well as the entire expression $F_i(\boldsymbol{\mu}, \lambda, \mathbf{n})$ are both equal to zero, since both sets of arguments should represent solutions. Solving equation (11.66) for $\boldsymbol{\mu}$ gives

$$\hat{\boldsymbol{\mu}}(\mathbf{n}) \approx \tilde{\boldsymbol{\mu}} - A^{-1} B(\mathbf{n} - \tilde{\mathbf{n}}), \tag{11.67}$$

where the $M + 1$ component of $\boldsymbol{\mu}$ refers to the Lagrange multiplier λ. The symmetric $(M + 1) \times (M + 1)$ matrix A is given by

$$A_{ij} = \begin{cases} \frac{\partial^2 \varphi}{\partial \mu_i \partial \mu_j}, & i, j = 1, \dots, M, \\ \frac{\partial^2 \varphi}{\partial \mu_i \partial \lambda} = -1, & i = 1, \dots, M, j = M + 1, \\ \frac{\partial^2 \varphi}{\partial \lambda^2} = 0, & i = M + 1, j = M + 1, \end{cases} \tag{11.68}$$

and the $(M + 1) \times N$ matrix B is

$$B_{ij} = \begin{cases} \frac{\partial^2 \varphi}{\partial \mu_i \partial n_j}, & i = 1, \dots, M, j = 1, \dots, N, \\ \frac{\partial^2 \varphi}{\partial \lambda \partial n_j} = 1, & i = M + 1, j = 1, \dots, N. \end{cases} \tag{11.69}$$

By using the error propagation formula (1.54), the covariance matrix for the estimators $U_{ij} = \text{cov}[\hat{\mu}_i, \hat{\mu}_j]$ is obtained from the covariance matrix for the data $V_{ij} = \text{cov}[n_i, n_j]$ by

$$\text{cov}[\hat{\mu}_i, \hat{\mu}_j] = \sum_{k,l=1}^{N} \frac{\partial \hat{\mu}_i}{\partial n_k} \frac{\partial \hat{\mu}_j}{\partial n_l} \text{cov}[n_k, n_l]. \tag{11.70}$$

The derivatives in (11.70) can be computed using (11.67) to be

$$\frac{\partial \hat{\mu}_i}{\partial n_k} = -(A^{-1} B)_{ik} \equiv C_{ik}, \tag{11.71}$$

where the matrices A and B are given by equations (11.68) and (11.69). What we will use here is not the entire matrix C, but rather only the $M \times N$ submatrix, excluding the row $i = M + 1$, which refers to the Lagrange multiplier λ. The final expression for the covariance matrix U can thus be expressed in the more compact form,

$$U = C V C^T. \tag{11.72}$$

The derivatives in (11.68) and (11.69) depend on the choice of regularization function and on the particular log-likelihood function used to define $\varphi(\boldsymbol{\mu}, \lambda)$ (11.40), e.g. Poisson, Gaussian ($\log L = -\chi^2/2$), etc. In the case, for example, where the data are treated as independent Poisson variables with covariance matrix $V_{ij} = \delta_{ij} \nu_i$, and where the entropy-based regularization function (11.54) is used, one has

$$\frac{\partial^2 \varphi}{\partial \mu_i \partial \mu_j} = -\alpha \sum_{k=1}^{N} R_{ki} R_{kj} \frac{n_k}{\nu_k^2}$$

$$+ \frac{1}{\mu_{\text{tot}}^2} \left[1 - \frac{\delta_{ij} \mu_{\text{tot}}}{\mu_i} + \log\left(\frac{\mu_i \mu_j}{\mu_{\text{tot}}^2}\right) + 2S(\boldsymbol{\mu}) \right] \tag{11.73}$$

and

$$\frac{\partial^2 \varphi}{\partial \mu_i \partial n_j} = \frac{\alpha R_{ji}}{\nu_j}. \tag{11.74}$$

The matrices A and B (and hence C) can be determined by evaluating the derivatives (11.73) and (11.74) with the estimates for $\boldsymbol{\mu}$ and $\boldsymbol{\nu}$ obtained in the actual experiment. Table 11.1 summarizes the necessary ingredients for Poisson and Gaussian log-likelihood functions. Note that for the Gaussian case, i.e. for the method of least squares, the quantities always refer to $\log L = -\frac{1}{2}\chi^2$, and not to χ^2 itself. The derivatives of Tikhonov and entropy-based regularization functions are given in Sections 11.5.1 and 11.5.2.

In order to determine the biases $b_i = E[\hat{\mu}_i] - \mu_i$, we can compute the expectation values $E[\hat{\mu}_i]$ by means of the approximate relation (11.67),

$$b_i = E[\hat{\mu}_i] - \mu_i \approx \tilde{\mu}_i + \sum_{j=1}^{N} C_{ij}(\nu_j - \tilde{n}_j) - \mu_i. \tag{11.75}$$

This can be estimated by substituting the estimator from equation (11.67) for μ_i and replacing ν_j by its corresponding estimator $\hat{\nu}_j = \sum_k R_{jk} \hat{\mu}_k$, which yields

$$\hat{b}_i = \sum_{j=1}^{N} C_{ij}(\hat{\nu}_j - n_j) = \sum_{j=1}^{N} \frac{\partial \hat{\mu}_i}{\partial n_j}(\hat{\nu}_j - n_j). \qquad (11.76)$$

The approximations used to construct $\hat{b}_i$ are valid for small $(\hat{\nu}_j - n_j)$, or equivalently, large values of the regularization parameter α. For small α, the matrix C in fact goes to zero, since the estimators $\hat{\mu}_i$ are then decoupled from the measurements n_j, cf. equation (11.71). In this case, however, the bias is actually at its largest. But since we will only use $\hat{b}_i$ and its variance in order to determine the regularization parameter α, the approximation is sufficient for our purposes.

By error propagation (neglecting the variance of the matrix C), one obtains the covariance matrix W for the $\hat{b}_i$,

$$W_{ij} = \text{cov}[\hat{b}_i, \hat{b}_j] = \sum_{k,l=1}^{N} C_{ik} C_{jl} \text{cov}[(\hat{\nu}_k - n_k), (\hat{\nu}_l - n_l)]. \qquad (11.77)$$

This can be computed by using $\hat{\nu}_k = \sum_m R_{km}\hat{\mu}_m$ to relate the covariance matrix $\text{cov}[\hat{\nu}_k, \hat{\nu}_l]$ to that of the estimators for the true distribution, $U_{ij} = \text{cov}[\hat{\mu}_i, \hat{\mu}_j]$, which is in turn related by equation (11.72) to the covariance matrix of the data by $U = CVC^T$. Putting this all together gives

$$\begin{aligned} W &= (CRC - C)\,V\,(CRC - C)^T \\ &= (CR - I)\,U\,(CR - I)^T, \end{aligned} \qquad (11.78)$$

where I is the $M \times M$ unit matrix. The variances $V[\hat{b}_i] = W_{ii}$ can be used to tell whether the estimated biases are significantly different from zero; this in turn can be employed as a criterion to determine the regularization parameter.

Table 11.1 Log-likelihood functions and their derivatives for Poisson and Gaussian random variables.

	Poisson	Gaussian (least squares)
$\log L$	$\sum_i (n_i \log \nu_i - \nu_i)$	$-\frac{1}{2}\sum_{i,j}(\nu_i - n_i)(V^{-1})_{ij}(\nu_j - n_j)$
$\frac{\partial \log L}{\partial \mu_i}$	$\sum_j \left(\frac{n_j}{\nu_j} - 1\right) R_{ji}$	$-\sum_{j,k} R_{ji}(V^{-1})_{jk}(\nu_k - n_k)$
$\frac{\partial^2 \log L}{\partial \mu_i \partial \mu_j}$	$-\sum_k \frac{n_k R_{ki} R_{kj}}{\nu_k^2}$	$-(R^T V^{-1} R)_{ij}$
$\frac{\partial^2 \log L}{\partial n_i \partial \mu_j}$	$\frac{R_{ij}}{\nu_i}$	$(V^{-1} R)_{ij}$

Before proceeding to the question of the regularization parameter, however, it is important to note that the biases are in general nonzero for all regularized unfolding methods, in the sense that they are given by some functions, not everywhere zero, of the true distribution. Their numerical values, however, can in fact be zero for particular values of μ. A guiding principle in unfolding is to choose a method such that the bias will be zero (or small) if μ has certain properties believed a priori to be true. For example, if the true distribution is uniform, then estimates based on Tikhonov regularization with $k = 1$ (11.42) will have zero bias; if the true distribution is linear, then $k = 2$ (11.43) gives zero bias, etc. If the true distribution is equal to a reference distribution $\mathbf{q}$, then unfolding using the cross-entropy (11.63) will yield zero bias.

11.7 Choice of the regularization parameter

The choice of the regularization parameter α, or equivalently the choice of $\Delta \log L$ (or $\Delta \chi^2$), determines the trade-off between the bias and variance of the estimators $\hat{\mu}$. By setting α very large, the solution is dominated by the likelihood function, and one has $\log L = \log L_{\max}$ (or with least squares, $\chi^2 = \chi^2_{\min}$) and correspondingly very large variances. At the other extreme, $\alpha \to 0$ puts all of the weight on the regularization function and leads to a perfectly smooth solution.

Various definitions of an optimal trade-off are possible; these can incorporate the estimates for the covariance matrix $U_{ij} = \text{cov}[\hat{\mu}_i, \hat{\mu}_j]$, the biases $\hat{b}_i$, and the covariance matrix of their estimators, $W_{ij} = \text{cov}[\hat{b}_i, \hat{b}_j]$. Here U and W will refer to the estimated values, $\hat{U}$ and $\widehat{W}$; the hats will not be written explicitly.

One possible measure of the goodness of the final result is the mean squared error, cf. equation (5.5), averaged over all bins,

$$\text{MSE} = \frac{1}{M} \sum_{i=1}^{M} (U_{ii} + \hat{b}_i^2). \tag{11.79}$$

The method of determining α so as to obtain a particular value of the MSE will depend on the numerical implementation. Often it is simply a matter of trying a value α, maximizing $\varphi(\mu, \lambda)$, and iterating the procedure until the desired solution is found.

One could argue, however, that the contribution to the mean squared error should be different for different bins depending on how accurately they are measured. Since the variance of a Poisson variable with mean value μ_i is equal to μ_i, one can define a weighted MSE,

$$\text{MSE}' = \frac{1}{M} \sum_{i=1}^{M} \frac{U_{ii} + \hat{b}_i^2}{\hat{\mu}_i}, \tag{11.80}$$

in analogy with the χ^2 used in the method of least squares. For Poisson distributed data, the quantity MSE$'$ represents the mean squared increase in the errors due to limited resolution. It is thus reasonable to require that this quantity be small.

A popular choice for the regularization parameter is based on the idea that, on average, each bin should contribute approximately one unit to the χ^2, i.e. α is determined such that $\chi^2 = N$. This can be generalized to the log-likelihood case as $\Delta \log L = \log L_{\max} - \log L = N/2$, since for Gaussian distributed $\mathbf{n}$ one has $\log L = -\chi^2/2$.

Naively one might expect that an increase in the χ^2 of one unit would set the appropriate level of discrepancy between the data $\mathbf{n}$ and the estimates $\hat{\boldsymbol{\nu}}$. This typically leads, however, to solutions with unreasonably large variance. The problem can be traced to the fact that the estimator $\hat{\nu}_i$ receives contributions not only from n_i but also from neighboring bins as well. The coupling of the estimators $\hat{\nu}_i$ to the measurements n_j can be expressed by the matrix

$$\frac{\partial \hat{\nu}_i}{\partial n_j} = \frac{\partial}{\partial n_j} \sum_{k=1}^{M} R_{ik} \hat{\mu}_k = (RC)_{ij}. \tag{11.81}$$

A modification of the criterion $\Delta \chi^2 = 1$ has been suggested in [Sch94] which incorporates this idea. It is based on an increase of one unit in an effective χ^2,

$$\Delta \chi_{\text{eff}}^2 = (\hat{\boldsymbol{\nu}} - \mathbf{n})^T RC V^{-1} (RC)^T (\hat{\boldsymbol{\nu}} - \mathbf{n}) = 1, \tag{11.82}$$

where the matrix RC effectively takes into account the reduced coupling between the estimators $\hat{\nu}_i$ and the data n_i.

Alternatively, one can look at the estimates of the biases and their variances. If the biases are significantly different from zero, then it is reasonable to subtract them. This is equivalent to going to a smaller value of $\Delta \log L$. As a measure of the deviation of the biases from zero, one can construct the weighted sum of squares,

$$\chi_b^2 = \sum_{i=1}^{M} \frac{\hat{b}_i^2}{W_{ii}}. \tag{11.83}$$

The strategy is thus to reduce $\Delta \log L$ (i.e. increase α) until χ_b^2 is equal to a sufficiently small value, such as the number of bins M. At this point the standard deviations of the biases are approximately equal to the biases themselves, and therefore any further bias reduction would introduce as much error as it removes.

The bias squared, the variance, and their sum, the mean squared error, are shown as a function of $\Delta \log L$ in Fig. 11.2. These are based on the example from Fig. 11.1, there unfolded by inverting the response matrix, and here treated using (a) maximum entropy and (b) Tikhonov regularization. The increase in the estimated bias for low $\Delta \log L$ reflects the variance of the estimators $\hat{b}_i$; the true bias decreases to zero as $\Delta \log L$ goes to zero. The arrows indicate solutions based on the various criteria introduced above; these are discussed further in the next section.

Further criteria for setting the regularization parameter have been proposed based on singular value analysis [Höc96], or using a procedure known as cross-validation [Wah79]. Unfortunately, the choice of α is still a somewhat open

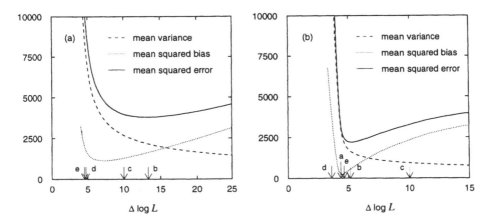

Fig. 11.2 The estimated mean variance, mean squared bias, and their sum, the mean squared error, as a function of $\Delta \log L$ for (a) MaxEnt and (b) Tikhonov regularization ($k = 2$). The arrows indicate the solutions from Figs 11.3 and 11.4: (b) is minimum MSE, (c) is $\Delta \log L = N/2$, (d) is $\Delta \chi^2_{\text{eff}} = 1$, and (e) is $\chi^2 b = M$. For the MaxEnt case, the Bayesian solution $\Delta \log L = 970$ is not shown. For Tikhonov regularization, (a) gives the solution for minimum weighted MSE.

question. In practice, the final estimates are relatively stable as the value of $\Delta \log L$ decreases, until a certain point where the variances suddenly shoot up (see Fig. 11.2). The onset of the rapid increase in the variances indicates roughly the natural choice for setting α.

11.8   Examples of unfolding

Figures 11.3 and 11.4 show examples based on maximum entropy and Tikhonov regularization, respectively. The distributions μ, ν and $\mathbf{n}$ are the same as seen previously in Figs 11.1(a)–(c), having $N = M = 20$ bins, all efficiencies ε equal to unity, and backgrounds β equal to zero. The estimators $\hat{\mu}$ are found by maximizing the function $\varphi(\mu, \lambda)$ (11.40), here constructed with a log-likelihood function based on independent Poisson distributions for the data. On the left, the original 'true' histograms μ are shown along with the unfolded solutions $\hat{\mu}_i$ and error bars $\sqrt{U_{ii}}$ corresponding to a given value of the regularization parameter α, or equivalently to a given $\Delta \log L$. On the right are the corresponding estimates of the biases b_i with their standard deviations $\sqrt{W_{ii}}$. These should not be confused with the true residuals $\hat{\mu}_i - \mu_i$, which one could not construct without knowledge of the true histogram μ. The estimates $\hat{b}_i$, on the other hand, are determined from the data.

Consider first Fig. 11.3, with the entropy-based regularization function $S(\mu) = H(\mu)$. Figure 11.3(a) corresponds to $\alpha = 1/\mu_{\text{tot}}$, i.e. the Bayesian prescription (11.59), and gives $\Delta \log L = 970$. We show this choice simply to illustrate that the prior density $\pi(\mu) = \Omega(\mu)$ does not lead to a reasonable solution. Although the standard deviations $\sqrt{U_{ii}}$ are very small, there is a large bias. The estimates $\hat{b}_i$ shown on the right are indeed large, and from their error bars one can see that they are significantly different from zero. Note that the estimated biases here are not, in fact, in very good agreement with the true residuals $\hat{\mu}_i - \mu_i$, owing to the approximations made in constructing the estimators $\hat{b}_i$, cf. equation (11.67). The approximations become better as $\Delta \log L$ is decreased, until the standard deviations $\sqrt{W_{ii}}$ become comparable to the biases themselves.

Figure 11.3(b) shows the result based on minimum mean squared error (11.79). This corresponds to $\Delta \log L = 13.3$ and $\chi_b^2 = 154$. Although the estimated biases are much smaller than for $\alpha = 1/\mu_{\text{tot}}$, they are still significantly different from zero.

The solution corresponding to $\Delta \log L = N/2 = 10$ is shown in Fig. 11.3(c). Here the biases are somewhat smaller than in the result based on minimum MSE, but are still significantly different from zero, giving $\chi_b^2 = 87$. In this particular example, requiring minimum weighted mean squared error (11.80) gives $\Delta \log L = 10.5$, and is thus similar to the result from $\Delta \log L = N/2 = 10$.

The results corresponding to $\Delta \chi_{\text{eff}}^2 = 1$ and $\chi_b^2 = M$ are shown in Figs 11.3(d) and (e), respectively. Both of these have biases which are consistent with zero, at the expense of larger variances compared to the results from $\Delta \log L = N/2$ or minimum MSE. The $\Delta \chi_{\text{eff}}^2 = 1$ case has $\chi_b^2 = 20.8$, and the $\chi_b^2 = M$ case has $\Delta \chi_{\text{eff}}^2 = 0.85$, so in this example they are in fact very similar.

Now consider Fig. 11.4, which shows examples based on the same distribution, but now using Tikhonov regularization with $k = 2$. The figures correspond to (a) minimum weighted MSE, (b) minimum MSE, (c) $\Delta \log L = N/2$, (d) $\Delta \chi_{\text{eff}}^2 = 1$, and (e) $\chi_b^2 = M$. Here in particular the solution from $\Delta \log L = N/2$ does not appear to go far enough; although the statistical errors $\sqrt{U_{ii}}$ are quite small, the biases are large and significantly different from zero ($b_i^2 \gg W_{ii}$). Reasonable results are achieved in (a), (b) and (e), but the requirement $\Delta \chi_{\text{eff}}^2 = 1$ (d) appears to go too far. The bias is consistent with zero, but no more so than in the case with $\chi_b^2 = M$. The statistical errors are, however, much larger.

A problem with Tikhonov regularization, visible in the right most bins in Fig. 11.4, is that the estimates can become negative. (All of the bins are positive only for Fig. 11.4(a).) There is in fact nothing in the algorithm to prevent negative values. If this must be avoided, then the algorithm has to be modified by, for example, artificially decreasing the errors on points where the negative estimates would occur. This problem is absent in MaxEnt unfolding, since there the gradient of $S(\boldsymbol{\mu})$ diverges if any μ_i approach zero. This penalty keeps all of the μ_i positive.

The techniques discussed in this chapter can easily be generalized to multidimensional distributions. For the case of two dimensions, for example, unfolding methods have been widely applied to problems of image restoration [Fri72, Fri80, Fri83, Ski85], particularly in astronomy [Nar86], and medical imaging [Lou92]. A complete discussion is beyond the scope of this book, and we will only illustrate some main ideas with a simple example.

Figure 11.5 shows an example of MaxEnt unfolding applied to a test photograph with 56×56 pixels. Figure 11.5(a) is taken as the 'true' image, representing the vector $\boldsymbol{\mu}$. In Fig. 11.5(b), the image has been blurred with a Gaussian resolution function with a standard deviation equal to 0.6 times the pixel size.

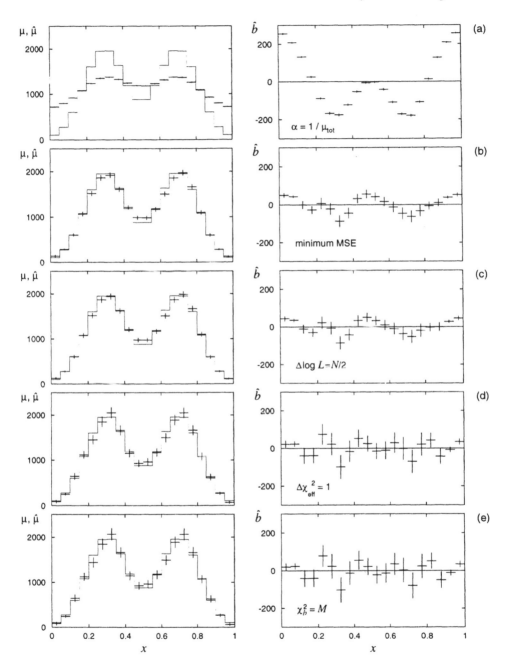

Fig. 11.3 MaxEnt unfolded distributions shown as points with the true distribution shown as a histogram (left) and the estimated biases (right) for different values of the regularization parameter α. The examples correspond to (a) the Bayesian prescription $\alpha = 1/\mu_{tot}$, (b) minimum mean squared error, (c) $\Delta \log L = N/2$, (d) $\Delta \chi^2_{eff} = 1$, and (e) $\chi^2_b = M$. In this example, the solution of minimum weighted MSE turns out similar to case (c) with $\Delta \log L = N/2$.

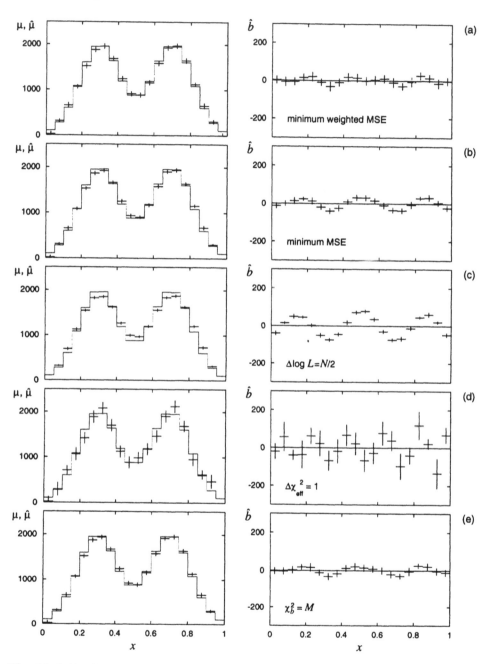

Fig. 11.4 Unfolded distributions using Tikhonov regularization ($k = 2$) shown as points with the true distribution shown as a histogram (left) and the estimated biases (right) for different values of the regularization parameter α. The examples correspond to (a) minimum weighted mean squared error, (b) minimum mean squared error, (c) $\Delta \log L = N/2$, (d) $\Delta\chi^2_{\text{eff}} = 1$, and (e) $\chi^2_b = M$.

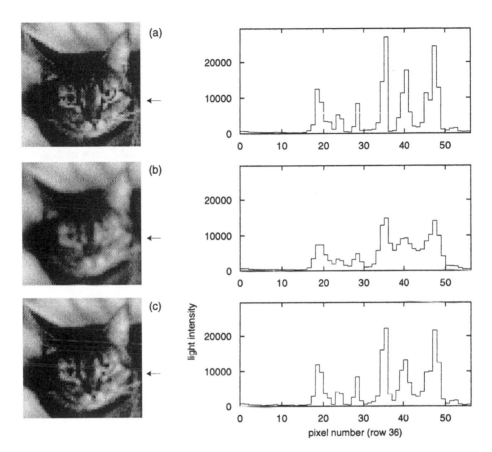

Fig. 11.5 (a) The original 'true' image μ. (b) The observed image n, blurred with a Gaussian point spread function with a standard deviation equal to 60% of the pixel size. (c) The maximum entropy unfolded image. The histograms to the right show the light intensity in pixel row 36 (indicated by arrows).

For purposes of this exercise, the effective number of 'photons' (or, depending on the type of imaging system, silver halide grains, photoelectrons, etc.) was assigned such that the brightest pixels have on the order of 10^4 entries. Thus if the number of entries in pixel i is treated as a Poisson variable n_i with expectation value ν_i, the relative sizes of the fluctuations in the brighter regions are on the order of 1% ($\sigma_i/\nu_i = 1/\sqrt{\nu_i}$). Figure 11.5(c) shows the unfolded image according to maximum entropy with $\Delta \log L = N/2$ where $N = 3136$ is the number of pixels. The histograms to the right of Fig. 11.5 show the light intensity in pixel row 36 of the corresponding photographs.

For this particular example, the method of maximum entropy has certain advantages over Tikhonov regularization. First, there is the previously mentioned feature of MaxEnt unfolding that all of the bins remain positive by construction. Beyond that, one has the advantage that the entropy can be directly generalized to multidimensional distributions. This follows immediately from the fact that the entropy $H = -\sum_j p_j \log p_j$ is simply a sum over all bins, and pays no attention to the relative values of adjacent bins. For Tikhonov regularization, one can generalize the function $S(\boldsymbol{\mu})$ to two dimensions by using a finite-difference approximation of the Laplacian operator; see e.g. [Pre92], Chapter 18.

A consequence of the fact that entropy is independent of the relative bin locations is that the penalty against isolated peaks is relatively slight. Large peaks occur in images as bright spots such as stars, which is a reason for MaxEnt's popularity among astronomers. For relatively smooth distributions such as those in Figs 11.3 and 11.4, Tikhonov regularization leads to noticeably smaller variance for a given bias. This would not be the case for distributions with sharp peaks, such as the photograph in Fig. 11.5.

A disadvantage of MaxEnt is that it necessarily leads to nonlinear equations for $\boldsymbol{\mu}$. But the number of pixels in a picture is typically too large to allow for solution by direct matrix inversion, so that one ends up anyway using iterative numerical techniques.

11.9 Numerical implementation

The numerical implementation of the unfolding methods described in the previous sections can be a nontrivial task. Finding the maximum of the function

$$\varphi(\boldsymbol{\mu}, \lambda) = \alpha \log L(\boldsymbol{\mu}) + S(\boldsymbol{\mu}) + \lambda \left[n_{\text{tot}} - \sum_{j=1}^{N} \nu_j \right] \tag{11.84}$$

with respect to $\boldsymbol{\mu}$ and the Lagrange multiplier λ implies solving the $M+1$ equations (11.64). If φ is a quadratic function of $\boldsymbol{\mu}$, then the equations (11.64) are linear. This occurs, for example, if one has a log-likelihood function for Gaussian distributed $\mathbf{n}$, giving $\log L = -\chi^2/2$, in conjunction with Tikhonov regularization. Methods of solution for this case based on singular value decomposition are discussed in [Höc96]. If φ contains, for example, a log-likelihood function based on the Poisson distribution, or an entropy-based regularization function, then the resulting equations are nonlinear and must be solved by iterative numerical techniques.

Consider as an example the case of a Poisson-based likelihood function, cf. equations (11.21), (11.22),

$$\log L(\boldsymbol{\mu}) = \sum_{i=1}^{N} (n_i \log \nu_i - \nu_i), \tag{11.85}$$

with the regularization function $S = H$ where H is the entropy (11.51).

A possible method of solution for MaxEnt regularization is illustrated in Fig. 11.6. The three axes represent three dimensions of $\boldsymbol{\mu}$-space, and the diagonal plane is a subspace of constant $\sum_i \nu_i = n_{\text{tot}}$. The two points indicated in the plane are the point of maximum entropy (all μ_i equal) and the point of maximum

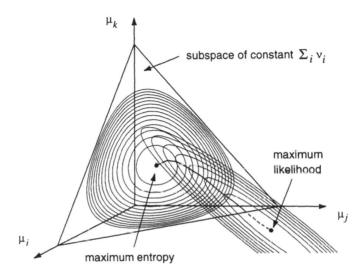

Fig. 11.6 Three dimensions of μ-space illustrating the numerical implementation of maximum entropy unfolding (see text).

likelihood. The curve connecting the two indicates possible solutions to (11.40) corresponding to different values of the regularization parameter α; for example, $\alpha = 0$ gives the point of maximum entropy; $\alpha \to \infty$ corresponds to the point of maximum likelihood. The curve passes through the points at which contours of constant entropy and constant likelihood touch. Note that the point of maximum likelihood is not in the allowed region with all $\mu_i > 0$. This is, in fact, typical of the oscillating maximum likelihood solution, cf. Fig. 11.1(d).

The program used for the MaxEnt examples shown in Figs 11.3 and 11.5 employs the following algorithm, which includes some features of more sophisticated methods described in [Siv96, Ski85]. The point of maximum likelihood usually cannot be used for the initial value of μ, since there one often has negative μ_i, and hence the entropy is not defined. Instead, the point of maximum entropy is taken for the initial values. This is determined by requiring all μ_i equal, subject to the constraint

$$\nu_{\text{tot}} = \sum_{i=1}^{N} \nu_i = \sum_{i=1}^{N}\sum_{j=1}^{M} R_{ij}\mu_j = \sum_{j=1}^{M} \varepsilon_j \mu_j$$

$$= n_{\text{tot}}, \tag{11.86}$$

where ε_j is the efficiency for bin j. The point of maximum entropy is thus

$$\mu_i = \frac{n_{\text{tot}}}{\sum_{j=1}^{M} \varepsilon_j}. \tag{11.87}$$

If one uses $S(\boldsymbol{\mu}) = \mu_{\text{tot}} H$, and if the efficiencies are not all equal, then the distribution of maximum $S(\boldsymbol{\mu})$ is not uniform, but rather is given by the solution to the M equations,

$$\log \frac{\mu_i}{\mu_{\text{tot}}} + \frac{S(\boldsymbol{\mu})\,\varepsilon_i}{n_{\text{tot}}} = 0, \quad i = 1, \ldots, M. \tag{11.88}$$

Starting from the point of maximum $S(\boldsymbol{\mu})$, one steps along the curve of maximum φ in the subspace of constant ν_{tot}. As long as one remains in this subspace, it is only necessary to maximize the quantity

$$\Phi(\boldsymbol{\mu}) = \alpha \, \log L(\boldsymbol{\mu}) + S(\boldsymbol{\mu}), \tag{11.89}$$

i.e. the same as $\varphi(\boldsymbol{\mu})$ but without the Lagrange multiplier λ, cf. (11.40). Simply requiring $\nabla \Phi = 0$ will not, however, lead to the desired solution. Rather, $\nabla \Phi$ must be first projected into the subspace of constant ν_{tot} and the components of the resulting vector set equal to zero. In this way the Lagrange multiplier λ never enters explicitly into the algorithm. That is, the solution is found by requiring

$$D\Phi = \nabla \Phi - \mathbf{u}(\mathbf{u} \cdot \nabla \Phi) = 0, \tag{11.90}$$

where $\mathbf{u}$ is a unit vector in the direction of $\nabla \nu_{\text{tot}}$. This is given by (cf. (11.10))

$$\frac{\partial \nu_{\text{tot}}}{\partial \mu_k} = \sum_{i=1}^{N} \sum_{j=1}^{M} R_{ij} \frac{\partial \mu_j}{\partial \mu_k} = \varepsilon_k, \tag{11.91}$$

so that the vector $\mathbf{u}$ is simply given by the vector of efficiencies, normalized to unit length,

$$\mathbf{u} = \frac{\boldsymbol{\varepsilon}}{|\boldsymbol{\varepsilon}|}. \tag{11.92}$$

We will use the differential operator D to denote the projection of the gradient into the subspace of constant ν_{tot}, as defined by equation (11.90).

One begins thus at the point of maximum entropy and takes a small step in the direction of $D \log L$. The resulting $\boldsymbol{\mu}$ is in general not directly on the curve of maximum Φ, but it will be close, as long as the step taken is sufficiently small. As a measure of distance from this curve one can examine $|D\Phi|$. If this exceeds a given limit then the step was too far; it is undone and a smaller step is taken.

If the resulting point $\boldsymbol{\mu}$ were in fact on the curve of maximum Φ, then we would have $\alpha \, D \log L + DS = 0$ and the corresponding regularization parameter would be

$$\alpha = \frac{|DS|}{|D \log L|}. \tag{11.93}$$

The parameter α can simply be set equal to the right-hand side of (11.93), and a side step taken to return to the curve of $D\Phi = 0$. This can be done using standard methods of function maximization (usually reformulated as minimization; cf. [Bra92, Pre92]). These side steps as well are made such that one remains in the subspace of $\nu_{\text{tot}} = n_{\text{tot}}$, i.e. the search directions are projected into this subspace. One then proceeds in this fashion, increasing α by means of the forwards steps along $D \log L$ and moving to the solution $D\Phi = 0$ with the side steps, until the desired value of $\Delta \log L = \log L_{\text{max}} - \log L$ is reached. Intermediate results can be stored and examined in order to determine the optimal stopping point.

Although the basic ideas of the algorithm outlined above can also be applied to Tikhonov regularization, the situation there is somewhat complicated by the fact that the solution of maximum $S(\boldsymbol{\mu})$ is not uniquely determined. For $k = 2$, for example, any linear function gives $S = 0$. One can simply start at $\mu_i = n_{\text{tot}}/M$ and set the regularization parameter α sufficiently large that a unique solution is found.

It is also possible with Tikhonov regularization to leave off the condition $\sum_i \nu_i = n_{\text{tot}}$, since here the regularization function does not tend to pull the solution to a very different total normalization. If the normalization condition is omitted, however, then one will not obtain exactly $\sum_i \nu_i = n_{\text{tot}}$. One can argue that $\hat{\nu}_{\text{tot}}$ should be an unbiased estimator for the total number of events, but since the bias is not large, the constraint is usually not included.

Bibliography

Among the following references, of special use for data analysis are the books by Barlow [Bar89], Brandt [Bra92], Eadie *et al.* [Ead71], Frodeson *et al.* [Fro79], Lyons [Lyo86], and Roe [Roe92], as well as the lecture notes by Hudson [Hud63, Hud64] and Yost [Yos85]. A collection of important statistical methods is included in the *Review of Particle Properties* by the Particle Data Group [PDG96], published every two years.

All80   W. Allison and J. Cobb, Relativistic Charged Particle Identification by Energy Loss, *Ann. Rev. Nucl. Part. Sci.* **30** (1980) 253.

Any91   V.B. Anykeyev, A.A. Spiridonov and V.P. Zhigunov, Comparative investigation of unfolding methods, *Nucl. Instrum. Methods* **A303** (1991) 350.

Any92   V.B. Anykeyev, A.A. Spiridonov and V.P. Zhigunov, Correcting factors method as an unfolding technique, *Nucl. Instrum. Methods* **A322** (1992) 280.

Arf95   George B. Arfken and Hans-Jurgen Weber, *Mathematical Methods for Physicists*, 4th edition, Academic Press, New York (1995).

Bab93   Wayne S. Babbage and Lee F. Thompson, The use of neural networks in γ–π^0 discrimination, *Nucl. Instrum. Methods* **A330** (1993) 482.

Bak84   Steve Baker and Robert D. Cousins, Clarification of the use of the chi-square and likelihood functions in fits to histograms, *Nucl. Instrum. Methods* **221** (1984) 437.

Bar89   R.J. Barlow, *Statistics: A Guide to the Use of Statistical Methods in the Physical Sciences*, John Wiley, Chichester (1989).

Bay63   T. Bayes, An essay towards solving a problem in the doctrine of chances, *Philos. Trans. R. Soc.* **53** (1763) 370. Reprinted in Biometrika, **45** (1958) 293.

Bel85   E.A. Belogorlov and V.P. Zhigunov, Interpretation of the solution to the inverse problem for the positive function and the reconstruction of neutron spectra, *Nucl. Instrum. Methods* **A235** (1985) 146.

Ber88   J.O. Berger and D.A. Berry, Statistical analysis and the illusion of objectivity, *Am. Sci.* **76**, No. 2 (1988) 159.

Bis95   Christopher M. Bishop, *Neural Networks for Pattern Recognition*, Clarendon Press, Oxford (1995).

Blo85   V. Blobel, Unfolding methods in high energy physics experiments, in *Proceedings of the 1984 CERN School of Computing*, CERN 85-09 (1985).

Bra92   S. Brandt, *Datenanalyse*, 3rd edition, BI-Wissenschaftsverlag, Mannheim (1992); *Statistical and Computational Methods in Data Analysis*, Springer, New York (1997).

CER97   CERN Program Library, CERN, Geneva, (1997). For software and documentation see `http://www.cern.ch/CERN/Computing.html` and links therein.

Cou92   Robert D. Cousins and Virgil L. Highland, Incorporating systematic uncertainties into an upper limit, *Nucl. Instrum. Methods* **A320** (1992) 331.

Cou95   Robert D. Cousins, Why isn't every physicist a Bayesian?, *Am. J. Phys.* **63** (1995) 398.

Cox46   R.T. Cox, Probability, frequency and reasonable expectation, *Am. J. Phys.* **14** (1946) 1.

Cra46   Harald Cramér, *Mathematical Methods of Statistics*, Princeton University Press, Princeton, New Jersey (1946).

Dag94   G. D'Agostini, On the use of the covariance matrix to fit correlated data, *Nucl. Instrum. Methods* **346** (1994) 306.

Dud73   Richard O. Duda and Peter E. Hart, *Pattern Classification and Scene Analysis*, John Wiley, New York (1973).

Dud88   Edward J. Dudewicz and Satya N. Mishra, *Modern Mathematical Statistics*, John Wiley, New York (1988).

Ead71   W.T. Eadie, D. Drijard, F.E. James, M. Roos and B. Sadoulet, *Statistical Methods in Experimental Physics*, North-Holland, Amsterdam (1971).

Fin74   Bruno de Finetti, *Theory of Probability: A Critical Introductory Treatment* (2 volumes), John Wiley, New York (1974).

Fis36   R.A. Fisher, The use of multiple measurements in taxonomic problems, *Ann. Eugen.* **7** (1936) 179; reprinted in *Contributions to Mathematical Statistics*, John Wiley, New York (1950).

Fis90   R.A. Fisher, *Statistical Methods, Experimental Design and Scientific Inference*, a re-issue of *Statistical Methods for Research Workers*, *The Design of Experiments*, and *Statistical Methods and Scientific Inference*, Oxford University Press, Oxford (1990).

Fri72   B.R. Frieden, Restoring with maximum likelihood and maximum entropy, *J. Opt. Soc. Am.* **62** (1972) 511.

Fri80   B.R. Frieden, Statistical models for the image restoration problem, *Comput. Graphics Image Process.* **12** (1980) 40.

Fri83   B.R. Frieden, *Probability, Statistical Optics, and Data Testing*, Springer, New York (1983).

Fro79   A.G. Frodesen, O. Skjeggestad and H. Tøfte, *Probability and Statistics in Particle Physics*, Universitetsforlaget, Oslo (1979).

Gna88   R. Gnanadesikan (ed.), *Discriminant Analysis and Clustering*, National Academy Press, Washington, DC (1988).

Gri86   Geoffrey Grimmett and Dominic Welsh, *Probability: an Introduction*, Clarendon Press, Oxford (1986).

Gri92   G.R. Grimmett and D.R. Stirzaker, *Probability and Random Processes*, 2nd edition, Clarendon Press, Oxford (1992).

Hel83   O. Helene, Upper limit of peak area, *Nucl. Instrum. Methods* **212** (1983) 319.

Her91   John Hertz, Anders Krogh and Richard G. Palmer, *Introduction to the Theory of Neural Computation*, Addison-Wesley, New York (1991).

Hig83   V.L. Highland, *Estimation of upper limits from experimental data*, Temple University Note COO-3539-38 (1983).

Höc96   Andreas Höcker and Vakhtang Kartvelishvili, SVD approach to data unfolding, *Nucl. Instrum. Methods* **A372** (1996) 469.

Hud63   D.J. Hudson, *Lectures on Elementary Statistics and Probability*, CERN 63-29 (1963).

Hud64   D.J. Hudson, *Statistics Lectures II: Maximum Likelihood and Least Squares Theory*, CERN 64-18 (1964).

Jam80   F. James, Monte Carlo theory and practice, *Rep. Prog. Phys.* **43** (1980) 1145.

Jam89   F. James and M. Roos, CERN Program Library routine D506 (long write-up), 1989; F. James, *Interpretation of the Errors on Parameters as given by MINUIT*, supplement to long write-up of routine D506 (1978).

Jam90   F. James, A review of pseudorandom number generators, *Comput. Phys. Commun.* **60** (1990) 329.

Jam91   F. James and M. Roos, Statistical notes on the problem of experimental observations near an unphysical region, *Phys. Rev.* **D44** (1991) 299.

Jam94   F. James, RANLUX: A Fortran implementation of the high-quality pseudorandom number generator of Lüscher, *Comput. Phys. Commun.* **79** (1994) 111.

Jay68   E.T. Jaynes, Prior probabilities, *IEEE Trans. Syst. Sci. Cybern.* **SSC-4** (1968) 227.

Jay86   E.T. Jaynes, Monkeys, kangaroos and N, in *Maximum Entropy and Bayesian Methods in Applied Statistics*, J.H. Justice (ed.), Cambridge University Press, Cambridge (1986) 26.

Jef48   Harold Jeffreys, *Theory of Probability*, 2nd edition, Oxford University Press, London (1948).

Koe84   K.S. Koelbig and B. Schorr, A program package for the Landau distribution, *Comput. Phys. Commun.* **31** (1984) 97.

Kol33   A.N. Kolmogorov, *Grundbegriffe der Wahrscheinlichkeitsrechnung*, Springer, Berlin (1933); English translation, *Foundations of the Theory of Probability*, 2nd edition, Chelsea, New York (1956).

Kul64   S. Kullback, *Information Theory and Statistics*, John Wiley, New York (1964).

Lan44   L. Landau, On the energy loss of fast particles by ionisation, *J. Phys. USSR* **8** (1944) 201.

Lec88   P.L. L'Ecuyer, Efficient and portable combined random number generators, *Commun. ACM* **31** (1988) 742.

Lee89   Peter M. Lee, *Bayesian Statistics: an Introduction*, Edward Arnold, London (1989).

Lin65   D.V. Lindley, *Introduction to Probability and Statistics from a Bayesian Viewpoint, Part 1: Probability, Part 2: Inference*, Cambridge University Press, Cambridge (1965).

Lön92   Leif Lönnblad, Carsten Peterson and Thorsteinn Rögnvaldsson, Pattern recognition in high energy physics with artificial neural networks – JETNET 2.0, *Comput. Phys. Commun.* **70** (1992) 167.

Lou92   A.K. Louis, Medical imaging: state of the art and future development, *Inverse Probl.* **8** (1992) 709.

Lüs94   Martin Lüscher, A portable high-quality random number generator for lattice field theory simulations, *Comput. Phys. Commun.* **79** (1994) 100.

Lyo86   L. Lyons, *Statistics for Nuclear and Particle Physicists*, Cambridge University Press, Cambridge (1986).

Lyo88   L. Lyons, D. Gibaut and P. Clifford, How to combine correlated estimates of a single physical quantity, *Nucl. Instrum. Methods* **A270** (1988) 110.

Mac69   H.D. Maccabee and D.G. Papworth, Correction to Landau's energy loss formula, *Phys. Lett.* **30A** (1969) 241.

Mar91   G. Marsaglia and A. Zaman, A new class of random number generators, *Ann. Appl. Probab.* **1** (1991) 462.

Mis51   Richard von Mises, *Wahrscheinlichkeit, Statistik und Wahrheit*, 2nd edition, Springer, Vienna (1951); *Probability, Statistics and Truth*, Allen and Unwin, London (1957).

Mis64   Richard von Mises, *Mathematical Theory of Probability and Statistics*, edited and complemented by Hilda Geiringer, Academic Press, New York (1964).

Mui82   R.J. Muirhead, *Aspects of Multivariate Statistical Theory*, John Wiley, New York (1982).

Mül95   Berndt Müller, Michael T. Strickland and Joachim Reinhardt, *Neural Networks: an Introduction*, 2nd edition, Springer, Berlin (1995).

Nar86   R. Narayan and R. Nityananda, Maximum entropy image restoration in astronomy, *Ann. Rev. Astron. Astrophys.* **24** (1986) 127.

Neu51   John von Neumann, Various techniques used in connection with random digits, *J. Res. NBS Appl. Math. Ser.* **12** (1951) 36; reprinted in *John von Neumann: Collected Works*, A.H. Taub (ed.), Vol. V, Pergamon Press, Oxford (1963).

Ney37   J. Neyman, Outline of a theory of statistical estimation based on the classical theory of probability, *Philos. Trans.*, **A 236** (1937) 333.

Oha94   A. O'hagan, *Kendall's Advanced Theory of Statistics*, Vol. 2B, *Bayesian Inference*, Edward Arnold, London (1994).

PDG96   The Particle Data Group, Review of particle properties, *Phys. Rev.* **D54** (1996) 1.

Per87   Donald H. Perkins, *Introduction to High Energy Physics*, Addison-Wesley, Menlo Park, California (1987).

Pet92   Carsten Peterson and Thorsteinn Rögnvaldsson, An Introduction to Artificial Neural Networks, in *Proceedings of the 1991 CERN School of Computing*, C. Verkerk (ed.), CERN 92-02 (1992).

Pet94   Carsten Peterson, Thorsteinn Rögnvaldsson and Leif Lönnblad, JET-NET 3.0 – A versatile artificial neural network package, *Comput. Phys. Commun.* **81** (1994) 185.

Phi62   D.L. Phillips, A technique for the numerical solution of certain integral equations of the first kind, *J. ACM* **9** (1962) 84.

Pre92   W.H. Press, B.P. Flannery, S.A. Teukolsky and W.T. Vetterling, *Numerical Recipes*, 2nd edition, Cambridge University Press, Cambridge (1992).

Roe92   Byron P. Roe, *Probability and Statistics in Experimental Physics*, Springer, New York (1992).

Sav72   Leonard J. Savage, *The Foundations of Statistics*, 2nd revised edition, Dover, New York (1972).

Sch94   Michael Schmelling, The method of reduced cross-entropy. A general approach to unfold probability distributions, *Nucl. Instrum. Methods* **A340** (1994) 400.

Sha48   C.E. Shannon, A mathematical theory of communication, *Bell Sys. Tech. J.* **27** (1948) 379, 623. Reprinted in C.E. Shannon and W. Weaver, *The Mathematical Theory of Communication*, University of Illinois Press, Urbana (1949).

Siv96   D.S. Sivia, *Data Analysis, a Bayesian Tutorial*, Clarendon Press, Oxford (1996).

Ski85   John Skilling and S.F. Gull, Algorithms and applications, in *Maximum Entropy and Bayesian Methods in Inverse Problems*, C. Ray Smith and W.T. Grady (ed.), D. Reidel, Dordrecht (1985) 83.

Ski86   John Skilling, Theory of maximum entropy image reconstruction, in *Maximum Entropy and Bayesian Methods in Applied Statistics*, J.H. Justice (ed.), Cambridge University Press, Cambridge (1986) 156.

Spr79   M.D. Springer, *The Algebra of Random Variables*, John Wiley, New York (1979).

Stu91   Alan Stuart and J. Keith Ord, *Kendall's Advanced Theory of Statistics*, Vol. I, *Distribution Theory*, 6th edition, Edward Arnold, London (1994); Vol. II, *Classical Inference and Relationships*, 5th edition, Edward Arnold, London (1991).

Tik63   A.N. Tikhonov, On the solution of improperly posed problems and the method of regularization, *Sov. Math.* **5** (1963) 1035.

Tik77   A.N. Tikhonov and V.Ya. Arsenin, *Solutions of Ill-Posed Problems*, John Wiley, New York (1977).

Wah79   G. Wahba, Smoothing and ill-posed problems, in *Solution Methods for Integral Equations*, Michael Goldberg (ed.), Plenum Press, New York (1979).

Yos85   G.P. Yost, *Lectures on probability and statistics*, Lawrence Berkeley Laboratory report LBL-16993 (1985).

Zec95   G. Zech, *Comparing statistical data to Monte Carlo simulation – parameter fitting and unfolding*, DESY 95-113 (1995).

Zhi83   V.P. Zhigunov, Improvement of resolution function as an inverse problem, *Nucl. Instrum. Methods* **216** (1983) 183.

Zhi88   V.P. Zhigunov, T.B. Kostkina and A.A. Spiridonov, On estimating distributions with the maximum entropy principle, *Nucl. Instrum. Methods* **A273** (1988) 362.

Index

lton Keynes UK
am Content Group UK Ltd.
W020636060724
93UK00007B/184